Steven Rose is Professor of Biology and Director of the Brain and Behaviour Research Group at the Open University. A biochemist by training, his research centres on the molecular and cellular mechanisms of memory formation, although he has also written and worked extensively on issues concerning the social framework and consequences of science. He is the author of many books, including, in Penguin, *The Chemistry of Life* (first published in 1966, now in its fourth edition), *Lifelines* (1997) and, with Richard Lewontin and Leon Kamin, *Not in Our Genes* (1984), and the editor of *From Brains to Consciousness?* (1998; Penguin, 1999). Steven Rose won the 1993 Rhône–Poulenc Science Book Prize for *The Making of Memory*.

Radmila Mileusnic was born in the former Yugoslavia in 1947, and received her first degree in biology at the Faculty of Natural Sciences and Mathematics, University of Belgrade. After taking post-graduate studies in Molecular Biology, she joined the Institute of Biochemistry at the School of Medicine and in 1978 obtained her Dr. Sc. In 1984 she was awarded an Alexander von Humboldt Fellowship and in 1991 became Professor of Biochemistry at the School of Medicine, University of Belgrade. She left Yugoslavia in 1993 and joined the Brain and Behaviour Research Group at the Open University, where she continues her two-decade long collaboration with Steven Rose. In 1995 she became a lecturer in Molecular Biology at the Open University.

STEVEN ROSE

WITH

RADMILA MILEUSNIC

THE
CHEMISTRY
OF LIFE

Fourth Edition

PENGUIN BOOKS

PENGUIN BOOKS

Published by the Penguin Group
Penguin Books Ltd, 27 Wrights Lane, London W8 5TZ, England
Penguin Putnam Inc., 375 Hudson Street, New York, New York 10014, USA
Penguin Books Australia Ltd, Ringwood, Victoria, Australia
Penguin Books Canada Ltd, 10 Alcorn Avenue, Toronto, Ontario, Canada M4V 3B2
Penguin Books (NZ) Ltd, Private Bag 102902, NSMC, Auckland, New Zealand

Penguin Books Ltd, Registered Offices: Harmondsworth, Middlesex, England

First published in Pelican Books 1966
Reprinted with revisions 1970
Second edition 1979
Third edition published in Penguin Books 1991
Fourth (revised) edition 1999
1 3 5 7 9 10 8 6 4 2

Set in 10/12 pt Postscript Monotype Plantin Light
Typeset by Strathmore Publishing Services
Printed in England by Clays Ltd, St Ives plc

Contents

For Hilary

who still demands that biochemistry be intelligible,
but also that it be modest and responsible enough
to know its own place in the scheme of things

Preface

The first edition of *The Chemistry of Life* was published in 1966. I had only a few years since completed a 'classical' biochemistry training at Cambridge, done a PhD in brain biochemistry in London and was completing a post-doctoral period in Rome when I wrote the book in 1963–4. I was very young and full of optimism at the time, a biochemical positivism which reflected the exuberant atmosphere of the advances in biochemical knowledge of the period. Such knowledge might not change the world, I thought, but it would help explain it, and I believed then, as I believe now, in the importance of opening up science and making it accessible to those outside the arcane elitism with which it is often surrounded. By the time the first fully revised edition appeared in 1979, I was more conscious of the problematic nature of the reductionist framework within which the book was set, and I had begun to find the relationship between my science and the social and philosophical context in which it is embedded more troubling and less self-evident than when I originally wrote it. These issues have concerned me increasingly over the years, and not least when I contemplated preparing this, the fourth, and (it may be) the final revision of the text during my research lifetime.

I never intended *The Chemistry of Life* as a textbook, but rather to convey some of the excitement I felt in my subject to a lay reader, by discussing its central concepts and methods without an overload of data and didactic presentation. However, perhaps by default and lack of alternatives, it soon found its way into schools and basic college biochemistry courses, and this use was extended during the 1970s when it became a set text for the original Open University Science Foundation Course. It is both flattering and rather chastening to have met over the past decades so many people who tell me that it was reading this book which persuaded them to study biochemistry at university. No author could deny hearing this with some pleasure, but I am never sure that those who made the choice in part as a result of reading it feel that it was the right one for them! In any event it has been a source of both surprise and happiness to me that a book I wrote first as a young post-doc should in this new incarnation be about to see my research career into its final years.

Of course, what counts as biochemistry has changed profoundly over these more than three decades. It isn't only that there has been at least a tenfold increase in the known biochemical 'facts' about the world, and several revolutions in research techniques, but that the conceptual framework within which those facts are organized and experiments conducted has been transformed. A biochemistry textbook which is written *ab initio* by a novice biochemist at the start of the 21st century would probably be organized rather differently from the way I originally arranged this one. Indeed, it might well turn out to be a molecular biology text instead. So with each new edition I have been confronted with the question of whether to retain something recognizably like the original format or to radically rethink it. I don't believe it is either innate conservatism or mere laziness which has led me to keep the original shape, but rather as a way of arguing that this remains a meaningful and accessible way of introducing biochemical ideas.

Biochemistry today lies in a mixed terrain between the expansionist claims of molecular biology and genetics, disciplines with a recognizably different philosophy and methodology despite their common inheritance, and the increasingly ordered understanding of cell biology, physiology and indeed evolutionary biology which together are unifying many areas of science which previously were seen as unrelated.

Further, while at the time of the first edition of the book the industrial uses of biochemistry were largely confined to fermentation technology and the increasing power of the pharmaceutical industry, today biotechnology and genetic engineering – terms which didn't even exist in the early 1960s – have become part of daily newspaper fare both for the claims they make to change the world we live in and for the social, economic and ethical dilemmas which their untrammelled development are seen to raise.

So, even despite the temptations of expanding in the direction of my own research interests in the biochemistry and cell biology of brain function, I have finally decided to keep the essential structure and mode of analysis of the original edition in place, with its central concern with the biochemistry of mammals, especially humans, and its themes of biochemistry as analysis, as metabolism and as control – very definitely therefore a biochemical and not a molecular or cell biological agenda. The treatment of these themes is of course greatly modified from the original, but the early chapters – from the basic chemical concepts of Chapter 1, the small and large molecules of Chapters 2 and 3, cell structure in 4, work, enzymes and metabolism in 5–9 – remain in the original sequence. The chapters on protein and nucleic acid synthesis and control mechanisms have come to occupy an increasing portion of the book, as has that on the biochemistry of immune mechanisms. The final chapter, which treats fairly lightly the issues raised by biotechnology and genetic engineering, could of course have made a book on its own, and I have had to exercise considerable restraint! And because my own range of biochemical knowledge has if anything contracted as biochemistry has itself expanded, I have again been heavily reliant on research colleagues in the revising. In particular, the revision, especially of Chapters 10 and 11, would not have been possible without my very dear colleague now of some twenty years' standing, Radmila (Buca) Mileusnic. The immunology chapter, 13, also reflects the considerable input of her partner, the distinguished immunologist Miroslav (Mirko) Simic. I am heavily indebted to them both. This new edition also contains a thoroughly revised and expanded set of illustrations. Once again I would like to thank Penguin Books, and my successive editors there, Ravi Mirchandani and Stefan McGrath, for their willingness to keep this old war-horse in harness.

January 1999
STEVEN ROSE

Acknowledgements

The following acknowledgements are made for the use of photographs: to Mr Tony King for Plates 1, 3, 6 and 7; to Dr D. H. Beaufay for Plate 2; to Heather Davies, Open University, for Plate 4; to Dr E. V. Kiseleva for Plate 5; and to Shelagh Rearden, Wye College, for Plate 8 (below).

INTRODUCTION

What is Biochemistry?

'Biochemistry is the study of the chemical constituents of living matter and of their functions and transformations during life processes' – the definition of biochemistry in almost any standard textbook.

One does not normally regard a kitchen as having much in common with a laboratory. Most sciences are associated with heavy and elaborate machinery, the creation of high temperatures, vast pressures, or the concentration of great amounts of energy. The prototype of biochemical experiments, however, is the frying of an egg. Material from a biological source (the chicken) is removed from extraneous surrounding substances (the shell), care being taken all the while not to disrupt its natural organized structure (by breaking the yolk). This partially purified biological material is then subjected, under carefully controlled and regulated conditions, to a number of mild chemical and physical treatments (the egg is gently heated in butter, and pepper and salt are added). Whether the ultimate product is fit for anything more than the dustbin depends entirely on the skill with which these separate operations are performed; the margin between a good breakfast and a charred and tasteless mess is very small.

This analogy is not as unlikely as might at first be imagined; it is very probable that most good biochemists would also make good cooks, for, like cooks, they deal with fragile substances derived from living or recently dead animals and plants, which they must handle rapidly, gently and subtly in order to obtain meaningful results. As in cooking, the type of operation performed is all-important; no one would mistake a fried for a boiled or a scrambled egg; the differences in procedure by which one is arrived at rather than another are small (heating in or out of the shell, with or without prior mixing), but they are critical to the end product. So with biochemistry. Small variations in experimental procedure, in the concentration of a reactant, in the acidity or alkalinity of the medium in which the reaction is being carried out, or in its temperature, cause critical changes to occur in the behaviour of the highly complex chemical entities being studied. Whether these behavioural alterations make sense or nonsense will depend on both the experimental and the theoretical ingenuity of the biochemist.

It is for this reason that biochemistry has been essentially a twentieth-century science. The wealth of sophisticated techniques and concepts needed to make an approach to the understanding of the chemical order and functioning of the living cell is such that they could not have arisen except on the secure foundations of more elementary sciences. In essence, the medieval doctors speculating on the composition of blood, or devotedly distilling retorts filled with urine, were performing biochemical operations. So was the Italian abbot, Lazzaro Spallanzani, who, in 1783, fed hawks with pieces of meat enclosed in wire boxes, trained the birds to vomit up the boxes at various subsequent times, and observed that the action of the gastric juices on the meat was progressively to liquefy it. He deduced that the liquefaction was the result of chemical interaction between certain substances in the gastric juices and the meat; such active principles were later, under the name of enzymes, recognized as the cornerstone of modern biochemistry.

Equally a biochemist was Friedrich Wohler, who, in 1828, synthesized the biological material urea from the non-biological cyanic acid and ammonia, thereby settling a debate as old as alchemy itself, by providing the first unassailable evidence that the substances present in living organisms are chemical entities which differ from those in the chemist's reagent bottles only in their complexity, and not by the introduction of any mysterious hypothesis of the 'nature of life'.

These were pioneer steps, and great strides forward were made in the hundred years between Wohler's synthesis of urea and the isolation of the first crystalline enzyme – also, by odd chance, one related to urea, 'urease' – by Sumner in America in

1926. But the really explosive growth of biochemistry had to wait on the consolidation of chemical theory, and the pushing forward of the frontiers of biology to a region where the distinction between it and 'chemical physiology' became obscure. By the 1930s, the time was at last ripe for the biochemists to take over. The first signs of the new biochemistry emanated from the German laboratories of Meyerhof and Warburg, which in the 1920s and early 1930s housed some of the most brilliant biochemists in the world. With the coming of the Nazis, many of these youngsters fled to England and America. They found refuge in the ever-hospitable laboratories of Frederick Gowland Hopkins at Cambridge, and in such American laboratories as that of the Rockefeller in New York. From this period, and until the recovery of continental European science and the rise of Japan in the late 1960s, dates the pre-eminence of British and American biochemistry.

From then on, the science of biochemistry expanded throughout West and East Europe, the US, Japan and the former USSR at an immense rate, more rapidly even than that of nuclear physics. From the 1960s on, dramatic breakthroughs in hitherto little-understood areas became so frequent as to be accepted by the practising biochemist almost with weary resignation, whilst, to cope with the tidal wave of learned papers and reports, publishers produce weekly issues of journals which only a short while ago were monthlies or even quarterlies. In order to ensure rapid distribution of their findings, some researchers have now abandoned the journals and instead use the electronic mail to circulate research findings to a charmed inner circle. Not content with this, biochemistry spills over into related disciplines, and biochemists have staged take-over bids for the journals of physiology and chemistry, to say nothing of such general magazines as *Nature* in Britain and *Science* in America. However, concurrently with its expansion, biochemistry began to fragment. Its practitioners began to refer to themselves as enzymologists, neurochemists, even mitochondriologists. To remain a plain biochemist became almost passé. The still newer science of molecular biology, derived from biochemistry and genetics, but increasingly their unruly child, contemptuous or impatient of the concerns of an older generation, now threatens to render mere biochemists redundant.

Yet despite the expansion and fragmentation of their subject, there has remained a common ethos amongst biochemists, a common method of approach to their problems that distinguished them from their colleagues in other, adjacent disciplines. A biochemist (at least, the sort this book is about) is not just a physiologist applying chemical tools to living things, nor an organic or physical chemist who is interested in the properties of the chemical substances of the cell. Both such types of scientist exist, and the biochemist will most likely get along very well with them; but at the same time they are recognizably different, asking different questions and demanding different standards from the answers they get. Biochemistry has arisen at the meeting point of many sciences, and has been fertilized and enriched by all, yet it is itself essentially unique.

Partly because of the spectacular intellectual and, increasingly, commercial success of molecular biology over the last four decades, and the relative lack of concern – if not downright scepticism – amongst many molecular biologists for some of the more traditional interests of biochemistry, a certain tension has grown up between the parent discipline and its offspring; memorably encapsulated in a phrase of the New York nucleic acid biochemist Erwin Chargaff in the aftermath of the non-biochemists Crick and Watson's success in determining the structure of DNA with all its genetic implications. 'Practicing biochemistry without a licence,' he called their work, and the taunt has been returned in kind over the years, if less epigrammatically. Far more than the biochemist's, the molecular biologist's commitment has been a reductionist belief that to understand DNA and its role in the synthesis of protein is to understand the biological world; all else, from the organized properties of cells to the workings of the brain and the mechanisms of evolution, will then fall into place.

A biochemist's concern – or at least, my sort of biochemist's concern – is somewhat more complex. Faced with a living cell, or the tissue or organ which is composed of several million of such cells in close proximity to one another, the types of questions we ask, and with which this book is primarily concerned, can be summarized under four major heads:

(1) What is the composition of the cell, in terms of individual chemical compounds which can be recognized as functionally different from one another and which can be separated by the techniques of chemical and physical fractionation?

(2) What are the relationships between these chemicals, and how are they made and converted one into another by the cell?

(3) How are these chemical interconversions controlled and regulated within the cell so as to enable it to maintain its organized structure and activities?

(4) What distinguishes the cell studied from those of other tissues, organs, or species? That is, in what way is its design related to the function that the cell performs within the living organism considered as a whole?

These questions, of course, all interconnect. Although they are arranged on the page in order of increasing complexity, it is not always necessary, or easier, to answer (1) before (2) or (3) before (4). An attempt to solve (2) may provide clues to (3) instead. But, nonetheless, a different approach is required for each. In fact, a case may be made that they are also those questions which have been asked, broadly speaking, at different stages in the development of biochemistry as a science. Thus, in the first phase of the history of biochemistry, the most important and pressing problem was the establishment of the nature and composition of the chemicals of the body. To this phase belong the work of Wohler and Sumner already referred to, and the massive development of the French and German schools of organic chemistry in the hands of such nineteenth-century giants as Berthelot, Liebig, and Fischer, which resulted in the identification and subsequent synthesis of most of the simpler chemicals utilized by living organisms. The approach to the biological macromolecules (those with molecular weights off 10 000 upwards to a million or more), such as proteins, fats, carbohydrates, and nucleic acids, required a different order of techniques to those available to earlier generations of chemists, despite the classical and painstaking analyses of such workers as Thudicum, who in 1884 listed in his *Treatise on the Chemical Constitution of the Brain* some 140 individual constituents, many of them complex combinations of fats, proteins, and carbohydrates, isolated by methods of extraction involving prolonged subjection of the tissue to conditions so extreme in acidity, alkalinity or temperature as to curl the hair of a present-day biochemist with horror. Indeed a long battle was fought, lasting until the 1920s, against those who believed that proteins and carbohydrates were really built of giant molecules, rather than merely loose associations, *colloids*, of smaller units.

The first detailed molecular structure and chemical sequence of a protein were worked out for the hormone insulin by Frederick Sanger in Cambridge in 1956, after nearly a decade of minute and laborious analysis which well deserved the Nobel Prize with which it was received; by the 1980s these procedures, along with those for determining the sequences of DNA and RNA, had become a routine piece of laboratory technique, scarcely worth raising an eyebrow when published in one of the biochemical journals. Now the greatest sequencing job ever, that of the entire length of human DNA – the human genome – estimated cost, $3 billion, and vast numbers of biochemist-years employed in the US, Japan and Europe, but essentially now a matter of routine, a glorified factory production line, is approaching completion. Important though many remaining problems may be, the great days of the age of biochemical analysis are now truly past. In this book, we discuss its findings in Chapters 1 to 3, before returning to some of the implications of such projects as that for the human genome at the very end.

Meanwhile, as the scale on which the biochemists could work has moved up, that on which physiologists and microscopists were working has moved down. The advent of the electron microscope for routine laboratory use in the 1950s united the two. It made possible magnifications of 100 000 times and more, and enabled the cell to be studied visually in great detail for the first time, so that biochemists could see in detail, as well as imagine, the inside of the cell whose properties they were investigating. As will be shown in Chapter 4, it became possible to deduce just where within the cell particular substances were located, and where particular reactions occurred. It became apparent that the cell was not merely a bag of randomly distributed chemicals, but that each substance had its own place and position. Today, the cell seems to resemble more closely the regularly patterned form of a complex, hierarchically ordered set of macromolecules than the primitive sack of 'protoplasm' that had been the nineteenth-century picture.

The second phase of biochemistry was one of *kinetics*, of drawing a route-map setting out the major pathways by which chemical transformations occur within the living cell, and of understanding at a molecular level the mechanism of each individual chemical reaction. A primitive experiment of this sort was Spallanzani's, already described. The recognition in the early nineteenth century of the phenomenon of *catalysis*, in which chemical reactions are assisted and accelerated by substances (catalysts) which are themselves unaltered during the reaction, led to the assumption by the Swede Berzelius, in 1836, that the vast range of chemical activities occurring in living tissues depended upon the existence of potent chemical catalysts within the cell. And so indeed it proved. The heated controversies which followed, and which found distinguished chemists lined up in bitter opposition over whether the chemical reactions characteristic of life could be performed in the test-tube in the absence of living organisms, were resolved by the brothers Buchner, who, in 1897, ground yeast with sand in a mortar and extracted from the mixture a distinctly dead juice which was nonetheless able satisfactorily to ferment sugar to produce alcohol.

The name *enzyme* was coined for the catalyst in the yeast juice that performed this desirable function, and the properties of this enzyme soon showed that it was protein in nature. As more and more catalysts were discovered and extracted from the cell, the name enzyme became accepted as a general one for the entire class of biochemical catalysts. It is now recognized that practically every chemical reaction that occurs within the body requires its specific enzyme to catalyse it. Each enzyme catalyses only a single reaction, and the complete synthesis or degradation of a complex substance – for example, the break-down of the starch in food to the sugar molecules of which it is composed during digestion in the gut, followed by the absorption of the sugar into the cells and its synthesis there to glycogen ('animal starch') or breakdown to carbon dioxide and water – requires a whole series of enzymes acting in sequence, one after the other. A reaction chain of this sort is called a metabolic pathway (Chapter 6). The mapping of these pathways, for the synthesis and breakdown of sugars, fats, and amino acids (a few of which are outlined in Chapters 8 and 9), was the work of the generation of biochemists of the 1930s, and the names of Krebs, Embden, Meyerhof, Warburg, and Dickens stand high in this respect.

One problem remained in the understanding of these metabolic chemical interconversions: that of the *energy-balance* of the reactions. Destructive reactions (sometimes called *catabolic*), such as those of the breakdown of the sugar, glucose to carbon dioxide and water, release considerable quantities of energy; synthetic (or *anabolic*) reactions, such as the manufacture of proteins or fats, are energy-requiring. It is necessary for the cell to strike a balance between energy-producing and energy-demanding reactions, for it cannot afford to run for long at either a profit of a loss. It was Fritz Lipmann, in New York, who showed, in 1941, that the cell runs a sort of energy-bank, which can trap and store the energy released by catabolism, and provide it again on demand for anabolism, and that this bank consists of the chemical adenosine triphosphate (ATP for short). The significance and properties of ATP, as they are at present envisaged, will be discussed in detail in Chapters 5 and 7.

The 1950s saw a change in emphasis from the analysis of biochemistry-as-kinetics to that of the biochemistry-as-information. The intellectual rationale for this transition was provided by the growth of the new sciences associated with

8

the development of computers with their theories of control, feedback and information flow. As more and more became known about the mechanisms of individual enzymic reactions, about their energy-requirements, and about the workings of series of enzymes in the harmony of metabolic pathways, biochemists seized on these new information concepts in order to probe the ways in which the cell controlled and regulated its own metabolism; how, so to speak, it decided at any one time how much glucose to break down to carbon dioxide and water, or how much new protein to synthesize. And the triumph of biochemistry-as-information was of course the spectacular solution to the problem of the mechanics of the accurate replication of giant molecules such as DNA and of the translation of the genetic messages coded for in the DNA into the structure of the proteins themselves, undoubtedly one of the key scientific developments of the twentieth century (Chapter 10). It was this development which led to a major branch of biochemistry becoming hived off under the new name of molecular biology, charismatically proclaimed by its founders, notably Francis Crick and James Watson, in the decades after their achievement in unravelling the structure of DNA, and whose ambivalent relationship with the rest of biochemistry has already been mentioned.

This picture of the cell as a self-regulating mechanism, continually changing, yet continually stable, is one of the most important and significant results of modern biochemistry (Chapters 11 and 12). In the mid-nineteenth century, the great French physiologist Claude Bernard had described the fundamental property of life as that of the ability to 'maintain the constancy of the internal environment'. Living organisms responded to exterior events impinging upon them in such a way as to absorb the effects of these events into their systems as rapidly, and with as little disturbance, as possible. They needed constantly to renew themselves, to recreate from within those portions of themselves destroyed in the rough-and-tumble of existence. The pattern of their bodies was fixed, although its individual components were forever changing. For this process, the name 'homeostasis' (staying the same) was invented in the 1920s. It is of course something of a misnomer, for a living organism, unlike self-regulating machines such as thermostats, does more than merely stay the same; it is born, grows, reproduces, ages and dies. In short it has a history, as an individual and as a member of a species. Perhaps a better word to describe how an organism can both maintain its internal environmental constancy *and* change, not merely in response to external environmental contingencies but also its own internal clock, would be *homeodynamic*, and I strongly recommend this alternative. Central to such homeodynamics are the mechanisms that all organisms, but particularly large multicellular ones such as mammals, maintain to protect themselves from external attack by toxic substances or predatory microorganisms; the biochemical workings of the *immune system* form the subject of Chapter 13.

It has become the task of today's biochemistry to transfer the concepts of homeostasis and homeodynamics from the body as a whole to the working of each individual cell within it. Only when this has been done does it become possible to define in chemical and physical terms how the organism as a whole, as the sum of its constituent cells, and all their myriad interactions with each other and with the external world, can function. Armed with such knowledge, we can go on to ask the most fundamental of all biochemical questions: 'What is life, and how did it arise?' (Chapter 14). One of the most extraordinary findings of a century of biochemistry has been the essential universality of biochemical processes observed in a multitude of diverse species from yeasts to seaslugs to humans. This universality, which emphasizes the continuity of living systems in a profoundly interconnected world, itself finds explanation through one of the great unifying principles of biology, that of evolution. But by stressing the many similarities between humans and other life-forms, it also raises quite sharply questions about the rights and duties of humans towards those life-forms.

The huge increase in our understanding of ourselves and

the mechanisms of our own functioning as living organisms, to say nothing of our understanding of how non-humans work, has come about by way of experiments made by biochemists, by physiologists and many other biological scientists, on other living creatures. Dissections of body parts, physiological operations upon them, and the extraction of chemicals from them, have formed an essential part of acquiring this knowledge. This knowledge has greatly contributed to human welfare by way of medical treatments, agricultural, and, today, biochemical engineering advances. Certainly such advances are seldom unequivocal; drugs have unwanted and unexpected effects, agricultural advances alter ecosystems in ways which are homeodynamic but unpredictable. Nonetheless, weighing the elimination of so many desperate and debilitating diseases and, however uneven, the extension of the human lifespan and growth of wealth, few would deny on which side the balance comes down. The price has been, and continues to be, the experimental study of non-human animals. Those proponents of 'animal rights' who argue that the discontinuities between animals and humans are so great that nothing of use to the human condition can be learned by such study are simply wrong; those who say that some animal experiments are unnecessary because substitutes can be found are partly right, but for many problems, such as the function of complex organs such as the brain, there can be no substitute. Those who argue that experiments may be trivial or unnecessary because many of the new products they are designed to test are not required are of course correct. Then the question becomes one of how the social priorities for our science should be determined. The argument that, just because of our very humanness we have particular duties towards non-human life-forms, and must treat them with respect, is one with which I would hope every biochemist would agree. But with those who would privilege 'animal rights' above those of humans, who charge those who work on animals in the interests of human welfare as 'speciesist', I profoundly disagree.

There, on this universal theme, earlier editions of this book might have ended. Today, however, there is yet more to biochemistry; a whole range of new techniques, of biotechnology and genetic engineering, have turned laboratory pursuits into major forces for industrial intervention. No account of the biochemistry of the 21st century would be complete without some description of the scale and significance of these developments, which form the substance of Chapter 15. Where once the task of the biochemist seemed at best to try to help explain the world, the technological revolution which began in the 1970s and 1980s has also begun to make it possible for biochemistry to change it. Whether for well or ill is part of the agenda confronting the world as we approach the millennium.

CHAPTER I

Before We Start

In this book we shall be describing the biochemical make-up and behaviour of the living organism in some detail. In order to do this, we need constantly to use certain chemical words, phrases, and ideas. To those who have studied chemistry, what follows here is familiar territory; they would do better to skip the rest of this chapter. But, for those who are unacquainted with its jargon, there are included here a few brief paragraphs of definition in the hope that, having been disposed of, they need not trouble us unduly later on.

Chemists work with substances, which they attempt to purify one from another by making use of differences in their physical properties. For example, a mixture of salt and sand is separated because salt dissolves in water and sand does not; later, the salt can be recovered by boiling off the water. A substance which cannot be split by such physical methods into separate components is a *compound*. All chemical compounds, and there are many hundreds of millions of them, are formed by combination, in varying proportions of two or more, of a small number (about a hundred) of chemical *elements*; the elements can neither be converted into each other nor split into simpler substances by chemical means. The elements are represented by symbols, C for carbon, O for oxygen, Na for sodium, etc.; the compounds are indicated by a combination of these symbols. For example, common salt – sodium chloride – is NaCl.

Atoms and molecules

The smallest particle of an element is an atom, although the atom itself has an internal structure made up of smaller particles, protons, neutrons and electrons. Neutrons and protons are packed together in the atomic nucleus. Neutrons are so named because they have no electrical charge, protons are positively charged, whilst electrons (sometimes written e^-) have an equal though opposite charge to the protons, making the atom neutral. The electrons are distributed between very precisely defined orbits or energy levels at varying distances away from the positive nucleus (they can be thought of like planetary orbits around the sun). Each of the energy levels is known as a valency shell, and there is a maximum number of electrons that can exist in each shell. Atoms can combine with each other in fixed proportions to form molecules and when this happens they do so in a way that attains this maximum number in their outermost valency shell, as this gives the atom greatest stability. The number of electrons a particular atom has to lose or gain to achieve this stability is described as the valency of the atom. There are two ways this gaining or losing of electrons can be done: either by transferring electrons between atoms, or by sharing them. For example, the sodium atom, Na, achieves stability by *losing* an electron, thus becoming positively charged. The Cl atom does it by *gaining* an electron to become negatively charged. Occasionally, two or more atoms of a single element may join together *covalently* to form a molecule of that element; thus the gases hydrogen (H) and oxygen (O) normally exist as molecules containing each two atoms, H_2, O_2. The terms *atomic weight* and *molecular weight* are used to express the relative weights of atoms and molecules compared to the weight of hydrogen, which is the lightest element and whose weight is arbitrarily defined as 1. Thus the atomic weight of oxygen is 16, meaning that the oxygen atom is 16 times as heavy as the hydrogen atom. The molecular weight of water (H_2O) is ($2 \times 1 + 16$) or 18. In general, the more complex a molecule, the larger its molecular weight. Salt (NaCl) has a molecular weight of 58, the sugar glucose ($C_6H_{12}O_6$) of 180, and some proteins of a million or more.

Ions, electrovalent bonds, and buffers

Charged atoms or molecules are called ions. If Na^+ and Cl^- are close together in solution, transfer of electrons between them easily occurs, and because of their opposing charges, they are mutually attracted, combining to form salt, NaCl. The bond between them is described as *electrovalent*. In solution in water, electrovalent compounds such as NaCl tend to separate into their component ions, and to join together again only in the solid material, thus:

$$NaCl \rightleftharpoons Na^+ + Cl^-$$

The reverse-headed arrows of this equation indicate that the reaction may, depending on circumstances, proceed in either direction. It is said to be *reversible*. When it is going from left to right, NaCl is said to *dissociate*; in the opposite direction, Na^+ and Cl^- are referred to as *associating* to form NaCl. Other reactions – for example the *oxidation* (addition of oxygen) of coal gas (carbon monoxide) to form carbon dioxide, which occurs when we light a gas fire – are *irreversible* under normal circumstances:

$$2CO + O_2 \longrightarrow 2CO_2$$

Compounds formed by the combination of two ions, such as NaCl, are called *salts*. Chemically, they are usually produced during the reaction of two other compounds, an *acid* and an *alkali*. Acids are substances containing the hydrogen ion, H^+ (for example hydrochloric acid, HCl). Alkalies, on the other hand, contain a negative ion, the *hydroxyl* ion OH^-, itself a combination between hydrogen and oxygen. An example is sodium hydroxide (caustic soda), NaOH. Acids and alkalies react in solution to give a salt plus water, thus:

$$NaOH + HCl \longrightarrow NaCl + H_2O$$

A further word should be said here about the concept of acidity and alkalinity, because although a full treatment of it properly belongs to text-books of physical chemistry, biochemists find they cannot get very far without taking it into consideration. An acid solution contains hydrogen ions, an alkaline one, hydroxyl ions. Both hydrogen and hydroxyl ions are generated by the ionization of water:

$$H_2O \rightleftharpoons H^+ + OH^-$$

This is a reversible reaction, and in reversible reactions the relative concentrations of the reactants always tend towards a constant equilibrium point. Therefore the amounts of hydrogen and of hydroxyl ions present in any aqueous solution must always bear a constant relation to one another. If we know the number of hydrogen ions present, the number of hydroxyl ions follows automatically. Thus an alternative way of regarding the *addition* of hydroxyl ions is to look at it as a *subtraction* of hydrogen ions from the solution. We can then define *both* acidity *and* alkalinity in terms of the concentration of hydrogen ions present in the solution; the more hydrogen ions, the more strongly acid the solution is. The hydrogen ion concentration of a solution is measured on a scale called the pH scale. The scale (which is logarithmic) runs from 0 to 14, 0 being the acid, and 14 the alkaline, end of the scale. Midway between the two, a pH of 7.0 represents neutrality.

When hydrogen ions are added to a solution, the pH decreases; when they are removed (or hydroxyl ions added), the pH rises. But when certain substances are present in the solution, they will tend to combine with any H^+ or OH^- ions added, and by removing them, they act so as to prevent the change in pH that would otherwise occur. Substances that act in this way are called *buffers*. In their ability to mop up hydrogen or hydroxyl ions, buffers act as regulators of pH. Such regulation is extremely important to the delicately balanced living cell, where sharp fluctuation in acidity and alkalinity can easily spell disaster.

Covalent compounds

By far the largest number of substances with which we shall have to deal are not salts at all but compounds of the element carbon. Carbon compounds are so universally distributed amongst living organisms, and are so numerous, that their study has been split from those of other chemicals under the special name of *organic chemistry*. The carbon atom attains stability most easily by sharing its electrons with four atoms of a monovalent element or two atoms of a divalent atom. The simplest organic compound is thus methane (marsh gas) which has the formula CH_4. It is often useful when discussing carbon compounds, though, to abandon the simpler notation of inorganic chemistry and to try to draw a picture of the molecule as it actually exists in space. Thus

is a two-dimensional drawing of the three-dimensional methane molecule, demonstrating that the carbon atom in methane is in fact entirely surrounded by hydrogen atoms. Really, though, even this is only an approximation to the actual three-dimensional structure of the molecule, which, if one had a microscope powerful enough to see it, would appear to be not flat at all, but pyramid-shaped:

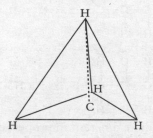

All the formulae we shall draw are, like that of methane, merely two-dimensional pictures of the real three-dimensional shape of the molecule.

In deference to the fact that the rule says that carbon must always be linked to four other atoms, four linking lines (*bonds*) are always drawn stretching out from any carbon atom. In cases such as that of carbon dioxide, CO_2, where the carbon atom is linked only to two atoms of oxygen, we draw the molecule as

$$O{=}C{=}O$$

showing that it is linked to oxygen not by *single* but by *double bonds*.

This type of bonding, by sharing electrons, is called *covalent*. Once made, covalent bonds are hard to break, and the substances which contain them therefore tend to be rather durable. In certain circumstances one partner atom of a covalent molecule may tend to obtain a greater or lesser share of the electrons than other partners. There is then a distribution of electric charge within the molecule which gives it a certain polarity, an important property, as we shall see in relation to macromolecules (electrovalent molecules are of course completely polarized).

Drawing pictures of the molecules in space also reveals certain other characteristics about them. For instance, when we examine the substances corresponding to the formula C_3H_6O, we find that two possible structures exist, represented by

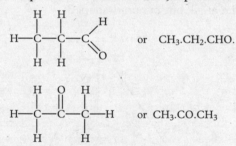

$$\text{or} \quad CH_3.CH_2.CHO.$$

$$\text{or} \quad CH_3.CO.CH_3$$

These two are quite different in properties: the first is *propionaldehyde*, formed by certain bacteria and a reactive, acrid-smelling substance; the second is the sweet-smelling *acetone*, familiar as the volatile base of nail varnish and aeroplane dope. The *structure* and *arrangement in space* of the molecules of organic compounds are thus critical to their behaviour. Substances (like propionaldehyde and acetone) whose overall chemical composition is the same but whose structure and shape are different are called *isomers*. *Isomerism* is a common occurrence amongst the chemicals of the living body, and we shall meet it frequently.

Yet another property of carbon atoms is their ability not only to link together in long, straight chains such as those we have been drawing, but also to form *branched chains*:

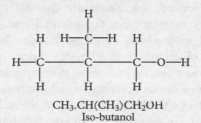

$CH_3.CH(CH_3)CH_2OH$
Iso-butanol

or even *rings*:

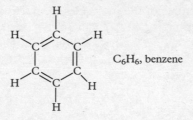

C_6H_6, benzene

Whenever we are dealing with molecules containing several carbon atoms, it is often useful to be able to refer to one in particular of the atoms. In order to do this, we *number* the carbon atoms in a molecule, starting at one end, preferably that nearest to some special group in the molecule, and going on to the other. Thus in *acetic acid*, we number the atoms starting with the acidic *carboxyl* group:

$$CH_3COOH$$

$$2 \quad 1$$

The advantage of such a numbering system will become obvious when we have to talk of larger, more complex molecules.

We shall discuss many more properties of carbon compounds during the course of this book, but however complicated they are, it is always the case that their structure and behaviour, from the simplest organic acid to the largest and most complex of proteins or nucleic acids, are governed by the same general rules that we have outlined here. There is nothing mysterious or inexplicable about the chemicals of the living organism. They may be very complicated and unstable and require special methods of handling, but the laws that they obey are those of chemistry and physics and no others.

With this preamble, we can move now straight to a discussion of the first of the sections into which we have divided biochemistry, that of the chemical nature of the substances that compose living cells, and which could be called *biochemistry as analysis* (Chapters 2 and 3).

CHAPTER 2

The Small Molecules

We regard the human body as solid enough. In everyday life, we prefer to know about its interior workings only what can be deduced by looking at its outside. The smooth skin of the perfect body is a delight, and we are content if we can trace underneath it the muscles of our arms or the pattern of our ribs. It requires some conscious effort to recollect, at least while we are healthy, that we have lungs, pancreas, liver, and kidneys, and that, dissected out, the organs that compose our body, that are our person, would not look so dissimilar to those of the animals we eat. Yet despite our distaste at the process ('Last week,' as Dean Swift remarked, 'I saw a woman flayed, and you will hardly believe how much it altered her person for the worse'), it is not beyond our imaginative powers.

The effort required by biochemists is greater. For them the body is not composed even of organs, but of an astronomical number of individual cells, each only visible under the microscope. And each cell, in its turn, must be regarded as composed of many distinct chemicals, all capable of being purified, crystallized, of having their molecular weight and structure determined, and finally, ultimate indignity, of being stored in a bottle on a shelf along with innumerable others of far less sacred origin. When biochemists visualize the body thus, they at once become conscious that despite the seeming dissimilarity between, say, liver and brain, or a snail and a human being, the common likenesses between the cells that compose them and the chemicals that make up the cells far outweigh the minor differences.

Against this reduction of the body to chemical terms a long and bitter struggle was waged – yet it was a crucial stage in the development of scientific understanding of the nature of life, quite apart from its importance in the development of medicine. The alchemists had known that if one concentrated urine one could crystallize out urea, and that the distillation of ants in a retort produced the pungent formic acid. And it had been known since antiquity that the fermentation and distillation of vegetable matter provided alcohol. Such chemicals, and many others, had been known long before the phase of research that we have called 'biochemistry as analysis' began. And when during the nineteenth century chemists began to transmute into biochemists, handling with increasing skill the always more complex materials they extracted from living tissues, the list became ever longer.

To determine the proportions of the various chemical elements present in the human body is comparatively simple. All that is needed is an incinerator, a balance, and the variety of techniques that analytical chemistry has developed to deal with the simpler problems presented by minerals. For instance, a flame photometer can recognize the metallic elements by the difference in the colour of the flame produced when a small sample of each is burnt in air. There are also specific ion electrodes that can quantify the amount of particular elements present in a solution.

Table 1 shows the results of such an analysis. Three elements, oxygen, hydrogen, and carbon, are present in overwhelming preponderance. The last figures in the table show why. The vast bulk of body-weight is provided by water, thus accounting for most of the hydrogen and oxygen.

The many other chemicals of the body are nearly all compounds of carbon, also with hydrogen and oxygen, sometimes with nitrogen, occasionally with sulphur. Most of the calcium and phosphorus is combined as calcium phosphate and forms the hard substance known as bone. But for all this it remains true that, as the biochemist–geneticist J. B. S. Haldane said, even the Archbishop of Canterbury is sixty-five per cent water.

The analysis of just how the elements present are combined is a more complex task, and one that is in no sense complete even today. It is possible, though, to describe the *classes* of compound present with some precision, even though not all is known about each individual member of the classes. The crudest, and primary, division may be illustrated by an experiment in which a piece of animal or plant tissue, such as

Table 1. *The composition of the human body by weight*

Class	Substance	% body weight
As elements	Oxygen	65
	Carbon	18
	Hydrogen	10
	Nitrogen	3
	Calcium	2
	Phosphorus	1.1
	Potassium	0.35
	Sulphur	0.25
	Sodium	0.15
	Chlorine	0.15
	Magnesium, Iron, Manganese, Copper, Iodine, Cobalt, Zinc }	Traces
As water and solid matter	Water	60–80
	Total solid material	20–40
As types of molecule	Protein	15–20
	Lipid	3–20
	Carbohydrate	1–15
	Small organic molecules	0–1
	Inorganic molecules	1

liver or leaf, is ground in a pestle and mortar with cold dilute acid and the resulting purée (or homogenate, as it is officially, if inaccurately, described) filtered to separate the soluble from the insoluble materials. The soluble fraction now contains nearly all the low-molecular-weight substances present in the tissue (those, that is to say, with a molecular weight of anything up to 10 000 or so), whilst the insoluble residue, typically whitish or pale brown in colour and rather pasty in consistency, contains the high-molecular-weight substances (with molecular weights of up to several millions), the proteins, nucleic acids and polysaccharides, as well as a variety of fatty substances.

For most biochemists, it is the giant molecules which are the really interesting and exciting ones, but even the most complex of giant molecules is built from, and its properties depend upon, a variety of smaller subunits, and it is these that we must begin by looking at. They include a number of inorganic ions, and up to a thousand or so different types of organic molecule, all of which fall into four broad classes: fatty acids; sugars; amino acids; and purines and pyrimidines.

LOW-MOLECULAR-WEIGHT COMPOUNDS

Soluble in the dilute acid of the experiment, and normally in solution also in the intact cell, are, first, a number of simple inorganic ions, notably the positively charged *potassium, sodium, calcium,* and *magnesium,* and the negatively charged *chloride* and *phosphate.* These ions are crucial for the functioning of the cell; they maintain its correct internal environment in the absence of which the macromolecules will be unable to function or may even disintegrate, and they are essential cofactors for many of the chemical reactions of the cell, as well as being directly involved in such cell functions as nerve transmission and muscular contraction.

Phosphates

The phosphate ion is an atom of phosphorus combined with four atoms of oxygen and has the formula PO_4^{3-}. It is significant because, unlike the other inorganic ions, it combines easily and enthusiastically with many organic compounds, due to its strong electronegativity. Because of its negativity it attracts protons (H^+ ions) very easily and in fact usually exists within the cell as orthophosphoric acid H_3PO_4:

but we will frequently shorten the formula, and write it simply as Ⓟ. It can readily form phosphate *salts* with, for example, sodium or potassium, or phosphate *esters* with organic alcohols. *Monoesters, diesters,* and *triesters* are theoretically possible:

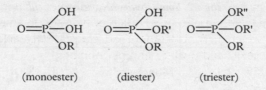

(monoester)   (diester)   (triester)

(where R stands for any organic alcohol) but in our extract we are likely to find only the monoesters and diesters. All the organic compounds which contain the alcoholic group CH_2OH can form such esters, and we shall come across *sugar* phosphates, *amino acid* phosphates, *hydroxyacid* phosphates, *amide* phosphates and *nucleoside* phosphates. The phosphate ester has the power of making an otherwise relatively inert organic compound biochemically very reactive indeed, and we shall find time and again, when the cell synthesizes or transforms an organic molecule, that the first step in the process is to convert it into its phosphate ester.

Phosphate groups can also combine, to give *di-* and even *triphosphates*, thus:

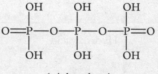

(triphosphate)

Such di- and triphosphate esters are common biologically, and are even more reactive than the monophosphate esters.

Organic acids

The simple carbon compounds listed in Table 2 are acids, although weak ones, because they readily ionize in water to give protons and a negatively charged group. The simplest is acetic acid and these acids can be purified from extracts of most living tissues. Acetic acid is of course in vinegar, and citric acid is found in citrus fruits such as lemons. They are chemically

Table 2. Some commonly occurring acids

Name	Type	Formula
Acetic	Monobasic	$CH_3.COOH$
Succinic	Dibasic	$CH_2.COOH$ | $CH_2.COOH$
Fumaric	Dibasic	$CH.COOH$ || $CH.COOH$
Lactic	Hydroxyacid	$CH_3.CHOH.COOH$
Malic	Hydroxyacid	$COOH$ | $CHOH$ | $CH_2.COOH$
Citric	Hydroxyacid	$COOH$ | CH_2 | $HO—C—COOH$ | CH_2 | $COOH$
Pyruvic	Keto acid	$CH_3.CO.COOH$
Oxaloacetic	Keto acid	$CO.COOH$ | $CH_2.COOH$

speaking not very exciting substances. Yet in life they are of great significance because of their ability to take part in complex chemical reactions with ease and speed. Starved of them, and of the sugars, the wonderfully complex machinery of protein and lipid that composes the cell comes tumbling down.

Fatty Acids

These acids are vitally important to the cell as they fulfil an important structural role combined as lipids in the cell membrane. The fatty acids are comparatively simple molecules containing unbranched hydrocarbon chains about 14–24 carbon atoms long, together with an acidic COOH group. The predominance of the long hydrocarbon over the relatively small acidic group means that the fatty acid is essentially nonpolar; that is, because of the symmetrical distribution of the electron shells there are no electrical charges. The fatty acids do not dissolve easily in water – they are *hydrophobic*. The hydrophobic nature of fatty acids has important implications for their role in membrane structure and the ways in which they interact. Another important property of the fatty acids is their degree of saturation; that is the number of carbon double bonds in their chain structure. Whilst single bonds in a molecule do not impose any constraint on its shape, because the parts of the molecule are free to rotate around them, double bonds are fixed and cannot be rotated. Thus two molecular configurations are possible for each double bond:

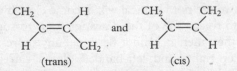

While in the *trans* position the symmetry of the groupings keeps the molecule straight, in the *cis* position the chain becomes bent and changes direction. Straight chain saturated fatty acids, and trans unsaturated acids, can pack together tightly, whilst kinked unsaturated ones cannot stack so neatly, and thus lipids formed from such acids are more fluid.

The acidic, polar group of the fatty acids is also important as it can combine easily with other organic molecules, as we shall see when we consider lipids later.

Sugars

Sugars can be written according to the general formula $C_nH_{2n}O_n$; in the simplest case, $n = 3$, and we can write two possible isomeric formulae:

CH₂OH.CHOH.CHO — glyceraldehyde (an aldose)

CH₂OH.CO.CH₂OH — dihydroxyacetone (a ketose)

Sugars which contain the aldehyde group ·CHO are *aldoses*, those which contain the ketone group CO are *ketoses*. If the formula for glyceraldehyde is drawn more fully, it can be seen, however, that it hides a further complexity. *Two* possible structures for glyceraldehyde can be drawn, which differ from one another in no other respect but that one is the mirror image of the other. These structures are *isomers* (see page 18):

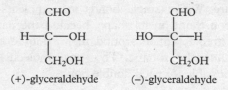

(+)-glyceraldehyde   (−)-glyceraldehyde

That these two are not identical will become clear if one imagines trying to rotate one of the drawings so as to superpose it directly on the other. It cannot be done, any more than mirror-images or right- and left-handed gloves can be superposed. They thus form a special class of isomers, known as *optical isomers*. Although these two molecules are chemically indistinguishable, their physical behaviour is not quite identical. They often form asymmetric, mirror-image crystals, and indeed Louis Pasteur, the discoverer of optical isomerism, first separated the two isomeric forms by picking out the different-shaped crystals. Also, and most important, they differ in the way they interact with plane-polarized light.*

We can draw the molecules of higher sugars, where n is greater than 3, by adding successive CHOH groups between carbon atoms 2 and 3 of glyceraldehyde of dihydroxyacetone. When this is done, it is found that each new carbon atom creates a new centre of asymmetry in the molecule, and can hence become the focus for further isomers. By the time $n = 6$, when the sugars are collectively known as *hexoses*, there are sixteen possible isomers for the aldoses and another sixteen for the ketoses. Fortunately, the cell is very discriminating. Half of each set of isomers, those based on (−) glyceraldehyde, never appear at all in nature, whilst of the others only a very few are at all common. Amongst the aldoses, the most important are glucose, galactose, and mannose. Amongst the ketoses we need only mention fructose which, though based on (+) glyceraldehyde, rotates plane-polarized light to the left.

$\overset{6}{C}HO$	$\overset{6}{C}HO$	$\overset{6}{C}HO$	$\overset{6}{C}H_2OH$
$\overset{5}{H}COH$	$\overset{5}{H}OCH$	$\overset{5}{H}COH$	$\overset{5}{C}=O$
$\overset{4}{H}OCH$	$\overset{4}{H}OCH$	$\overset{4}{H}OCH$	$HOCH$
$\overset{3}{H}OCH$	$\overset{3}{H}COH$	$\overset{3}{H}COH$	$\overset{3}{H}COH$
$\overset{2}{H}COH$	$\overset{2}{H}COH$	$\overset{2}{H}COH$	$\overset{2}{H}COH$
$\overset{1}{H}_2COH$	$\overset{1}{H}_2COH$	$\overset{1}{H}_2COH$	$\overset{1}{H}_2COH$
(+)-galactose	(+)-mannose	(+)-glucose	(+)-fructose

*In particular, when plane-polarized light is passed through a solution of one of the two isomers of glyceraldehyde, it is found that the plane of polarization of the light is rotated; each of the two mirror-images, however, rotates the plane in opposite directions. Viewed through one isomer the plane of the light is rotated to the right (+), whilst through the other it is rotated to the left (−). This power of rotating the plane of polarization of light is common to all molecules which are, like glyceraldehyde, asymmetric – that is, can be drawn as non-identical mirror-images.

The hexoses have the important property of being able to form ring-type structures; drawn diagrammatically, the process of ring-formation can be shown like this:

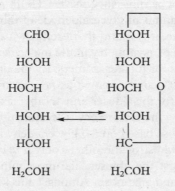

(the two molecules have identical formulae). But a better way of showing the ring is like this:

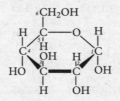

We shall in future frequently represent the glucose molecule this way. Dissolved in water, nearly all the glucose is in this closed-ring form (99.976 per cent to be precise). Drawn like this, the nature of the asymmetry of the CHOH groups can be seen more clearly; some of the OH groups stick up above the plane of the ring, whilst the rest fall below the plane. All the other hexose isomers have a slightly different pattern of 'above-and-belowness' of the OH groups.

This however creates another asymmetric C atom at C1. The OH group on C1 can be either *below* or *above* the plane of the ring, in the α or β position. This type of stereoisomerism is important because it is C1 that is the starting point for the

polymerization of sugars into polysaccharides. Thus the particular configuration at this atom can affect the whole shape of the
polysaccharide chain and hence its biological function.

A similar type of ring-closure occurs with the ketose sugars
such as fructose, but in this case the ring is a five-membered
one, thus:

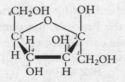

The sugars can combine with one another in their ring
forms to produce *chains* of linked sugar units. Thus carbon 1 of
one molecule links, via an oxygen atom, with one of the other
carbons of a second molecule, generally carbon 4. Such a bond
is known as a *glycosidic linkage*. An example is the compound
between one molecule of glucose and one of fructose:

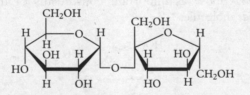

Here the bond is between C1 and C4, and in the α position, so
the molecule is 1—4α, and the substance is better known as
sucrose – common table sugar. Similarly, two glucose
molecules may combine to form *maltose*:

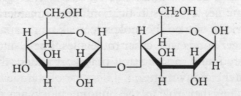

This process of linking (polymerization) can be repeated indefinitely, to build up chains of sugar units. Such chains, which may be several hundreds of units long, are *carbohydrates* (polysaccharides), the first of the giant molecules we shall shortly discuss.

Although the most important sugars are those containing 6 carbon atoms ($n = 6$), there are some examples of 7-carbon sugars, and rather more of 4- and 5-carbon sugars. When $n = 5$, the sugars are called pentoses, the most important of them being *ribose*. Like the hexoses, the pentoses normally exist in ring form and ribose can be drawn like this:

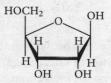

2-deoxyribose, in which the carbon atom at position two has lost its oxygen, also occurs. We shall shortly meet both ribose and deoxyribose as important constituents of the complex *nucleotide* molecules.

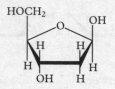

More complicated sugars exist because of the capacity for substituting ring hydroxyls with various other groups, e.g. NH_2, to give galactosamine, or $NHCOCH_3$ to give N-acetyl-galactosamine. These substitutions can considerably change the electric charge on the molecule, making it capable of forming stronger bonds with other molecules. These sugars are most commonly found in structural polysaccharides where stability and greater rigidity matter.

As well as their role as the building blocks for polysaccharides, the importance of the sugars biologically lies in the fact that they are the normal source of most of the energy utilized by the body; it is by the oxidation of glucose that the cell obtains the energy it needs for all the rest of its activities.

Amino acids

The amino acids are the building blocks of which the giant protein molecules are composed. Each amino acid contains nitrogen as well as carbon, hydrogen, and oxygen, and their general formula can be written:

$$\begin{array}{c} NH_2 \\ | \\ R\!-\!C\!-\!COOH \\ | \\ H \end{array}$$

where R may be any one of a number of different groups; in the simplest case, R is hydrogen and the amino acid is *glycine*:

$$\begin{array}{c} NH_2 \\ | \\ H\!-\!C\!-\!COOH \\ | \\ H \end{array}$$

There are about twenty naturally occurring amino acids in all; each, when pure, is a whitish powder with a faint but distinctive smell. The formulae of some of the more important of them are listed in Table 3. As with the sugars, isomeric forms of the amino acids can exist, as mirror-images of one another, thus:

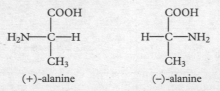

$$\begin{array}{ccc} COOH & \qquad & COOH \\ | & & | \\ H_2N\!-\!C\!-\!H & & H\!-\!C\!-\!NH_2 \\ | & & | \\ CH_3 & & CH_3 \end{array}$$

(+)-alanine   (–)-alanine

Also, as with the sugars, only one form is the naturally occurring one; in this case the one based on (–)-glyceraldehyde. The other isomers are neither produced nor utilized by animal cells, though they are used by some bacteria.

Table 3. Common amino acids found in proteins

Name	Abbreviation	Symbol	Formula
Alanine	Ala.	A	$CH_3-\overset{\overset{H}{\mid}}{\underset{\underset{+}{NH_3}}{C}}-COO^-$
Arginine	Arg.	R	$H_2N-\overset{\mid}{\underset{\underset{+}{NH_2}}{C}}-NH-CH_2-CH_2-CH_2-\overset{\overset{H}{\mid}}{\underset{\underset{+}{NH_3}}{C}}-COO^-$
Asparagine	As.	N	$\overset{NH_2}{\underset{O}{C}}-CH_2-\overset{\overset{H}{\mid}}{\underset{\underset{+}{NH_3}}{C}}-COO^-$
Aspartic acid	Asp.	D	$\overset{NH_2}{\underset{O}{C}}-CH_2-\overset{\overset{H}{\mid}}{\underset{\underset{+}{NH_3}}{C}}-COO^-$
Cysteine	Cyst.	C	$HS-CH_2-\overset{\overset{H}{\mid}}{\underset{\underset{+}{NH_3}}{C}}-COO^-$
Glutamic acid	Glut.	E	$\overset{^-O}{\underset{O}{C}}-CH_2-CH_2-\overset{\overset{H}{\mid}}{\underset{\underset{+}{NH_3}}{C}}-COO^-$
Glutamine	Glun.	Q	$\overset{NH_2}{\underset{O}{C}}-CH_2-CH_2-\overset{\overset{H}{\mid}}{\underset{\underset{+}{NH_3}}{C}}-COO^-$
Glycine	Gly.	G	$H-\overset{\overset{H}{\mid}}{\underset{\underset{+}{NH_3}}{C}}-COO^-$
Histidine	His.	H	$HC=C-CH_2-\overset{\overset{H}{\mid}}{\underset{\underset{+}{NH_3}}{C}}-COO^-$
Isoleucine	Ile.	I	$CH_3-CH_2-\overset{\mid}{\underset{CH_3}{CH}}-\overset{\overset{H}{\mid}}{\underset{\underset{+}{NH_3}}{C}}-COO^-$
Leucine	Leu.	L	$\overset{CH_3}{\underset{CH_3}{CH}}-CH_2-\overset{\overset{H}{\mid}}{\underset{\underset{+}{NH_3}}{C}}-COO^-$
Lysine	Lys.	K	$H_3\overset{+}{N}-CH_2-CH_2-CH_2-CH_2-\overset{\overset{H}{\mid}}{\underset{\underset{+}{NH_3}}{C}}-COO^-$

Table 3 (continued). Common amino acids found in proteins

Name	Abbreviation	Symbol	Formula
Methionine	Met.	M	CH_3—S—CH_2—CH_2—$\overset{\overset{H}{\mid}}{\underset{\underset{+}{NH_3}}{C}}$—$COO^-$
Phenylalanine	Phen.	F	CH_2—$\overset{\overset{H}{\mid}}{\underset{\underset{+}{NH_3}}{C}}$—$COO^-$
Proline	Prol.	P	H_2C, H_2C ring with N—H, C—COO^-, H_2
Serine	Ser.	S	HO—CH_2—$\overset{\overset{H}{\mid}}{\underset{\underset{+}{NH_3}}{C}}$—$COO^-$
Threonine	Thre.	T	CH_3—$\overset{\overset{OH}{\mid}}{\underset{\underset{}{H}}{C}}$—$\overset{\overset{H}{\mid}}{\underset{\underset{+}{NH_3}}{C}}$—$COO^-$
Tryptophan	Trp.	W	C—CH_2—$\overset{\overset{H}{\mid}}{\underset{\underset{+}{NH_3}}{C}}$—$COO^-$, CH, N—H (indole ring)
Tyrosine	Tyr.	Y	HO— ring —CH_2—$\overset{\overset{H}{\mid}}{\underset{\underset{+}{NH_3}}{C}}$—$COO^-$
Valine	Val.	V	$\overset{CH_3}{\underset{CH_3}{}}CH$—$\overset{\overset{H}{\mid}}{\underset{\underset{+}{NH_3}}{C}}$—$COO^-$

Inspection of the formula of the amino acids shows that they contain both an acidic group (the carboxyl-group, COOH) and an alkaline group (the amino-group, NH_2). In a neutral solution both of these groups are ionized, and it would be more correct to write the formula of an amino acid in water as:

$$R—\overset{\overset{NH_3^+}{\mid}}{\underset{\underset{H}{\mid}}{C}}—COO^-$$

Depending on the nature of R and the pH of the solution, amino acids may be positively or negatively charged, or neutral. The presence of different R groups can also considerably affect the electrical charge on the molecule, making the amino acid predominantly electropositive or electronegative, e.g. the carboxyl side groups of aspartate and glutamate make them negative ions whilst lysine and arginine have positively charged NH_3^+ groups. The ionic properties of the amino acids means that they can act as buffers (page 16) in solution. Amino acids can combine with one another to produce chains of linked units. The peptide bond is formed by joining the amino group of one amino acid to the carboxyl of a second. This process of linking can be repeated, to build up chains of units. Such chains, which may be several hundreds of units long, are proteins. The number, nature and size of amino acid residues (often abbreviated to just the letter R) in each chain determine its folding into higher order structures such as secondary and tertiary structures, as we shall see in the next chapter.

Purines and pyrimidines

Just as amino acids are the starting points for the proteins, so another group of giant molecules, the nucleic acids, find their building blocks in substances derived from two simple related ring compounds, *purine* and *pyrimidine*.

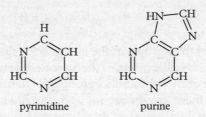

pyrimidine   purine

The most frequently occurring derivatives of pyrimidine are *cytosine*, *uracil*, and *thymine*, whilst the commonest purines are *adenine* and *guanine*. These so-called *bases* differ from each other because of the different groups substituted into the rings. It is these substitutions, together with the shape and spatial orientation of each of the rings, that make possible the specificity of base pairing and determine the physical characteristics of the nucleic acids, as we shall see in the next chapter.

Mostly the purines and pyrimidines do not occur free, but are combined with sugar, *ribose*. Generally, one or more phosphate groups are also present. Such molecules, containing purines or pyrimidines, ribose and phosphate, are known as *nucleotides*, thus emphasizing the part they play in the formation of the nucleic acids (see page 68).

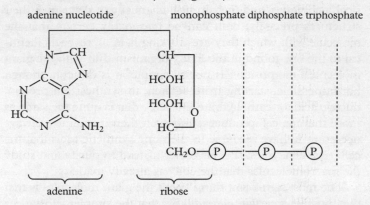

Depending on the number of phosphate groups present (it will be remembered that we have already shown how phosphate groups can readily be converted into polyphosphate groups), nucleotide mono-, di-, and triphosphates are possible. As was mentioned previously, the addition of phosphate groups to a substance has the effect of conferring biological reactivity on it, and, in conformity with this rule, the nucleotides, and especially their triphosphates, are amongst the most highly reactive of any the biochemist has to deal with, and they are utilized by the cell in many important reactions, especially biosynthetic mechanisms. Of all the chemicals to which we shall refer, none will be more frequently mentioned than *adenine nucleotide triphosphate*, which we shall henceforward abbreviate to ATP.

CHAPTER 3

Macromolecules

The low-molecular-weight compounds that we have described occupy an intermediate zone between organic chemistry and biochemistry. Given time and the right optical isomers to start with, the organic chemist can synthesize most of them by classical techniques, and the physical chemist can determine their structures precisely; both can be reasonably certain that the molecule with which they are working is in all respects identical to the one found in the living organism. But with the giant molecules, the position is not so simple; it is difficult, though not impossible, starting from scratch, to synthesize a protein, nucleic acid, or carbohydrate molecule that is quite the same as those that the cell produces, and even then it is almost always necessary to have recourse to the same synthetic tools that the cell uses – enzymes. It is far easier instead to purify and study the macromolecules that the cell has already made.

The most significant thing about the giant molecules is that they possess a certain *individuality* that the simpler substances lack. Two molecules of glucose, or of ATP, are identical from whatever source they are prepared, in the same way as are two molecules of copper sulphate or water. But within the long chains of linked repeating units that comprise the macromolecules there is room for variety and permutations of great subtlety, enough to puzzle chemists and their classical armoury of analytical tools. Indeed until the 1930s a great debate raged as to whether the macromolecules could ever properly be studied in their own right, as the earlier biochemists believed them to be simply physical aggregates of the smaller molecules that we have already discussed.

However, what is confusing to the analytical scientist is much less so for the cell, whose precision in distinguishing tiny variations in molecules whose weight may be in the order of millions is as great as the analytical power of the best of computers. The classical case is that of haemoglobin, a blood protein which contains some three hundred amino acids; changing the sequence of the chain, by swapping just one of these amino acids for another, is sufficient to prevent the haemoglobin from functioning properly, and results in the disease known as sickle-cell anaemia. We shall meet many similar cases, for such specificity of structure is one of the major differences between living and non-living things.

The giant molecules are the stuff of life; our muscles, skin and hair are protein fibres; beneath our skin, giant molecules of lipid are laid down in layers of subcutaneous fat; around us, the bark of trees and the stems of plants are long, close-packed molecules of the carbohydrate cellulose. The feel of a living thing is the feel of a complex meshwork of interacting macromolecules.

The chemist puts them firmly in their place, defining them briefly as substances with molecular weights of 10 000 and upwards, with somewhat indeterminate physical properties, composed by the polymerization (joining together in long chains) of simple low-molecular-weight units. But beyond saying this, chemists tend to regard them as rather outside their domain. So we have come at last to the realm of biochemistry proper.

Biochemical methods of dealing with the macromolecules tend to be somewhat *ad hoc*; biochemists have rather to take them as they find them. So we will increasingly find it necessary to abandon something of the classical outlook of the chemist. It is difficult to ask of a macromolcule 'is it a pure substance?', for one has few criteria of purity. One extracts the molecule from the living tissue, and in doing so is forced to disrupt the delicately balanced unity of the cell, to submit it to acids, salt solutions, organic solvents, ion-binding resins, and many other such chemical probes. When we finally have left a substance which we cannot break down any further without doing something really drastic, like hydrolysing* or oxidizing it,

---

*'Hydrolysis' means the splitting of a molecule by adding the elements of water, H_2O, to it. It is generally carried out by heating the substance in solution with dilute acid or alkali.

and which we cannot resolve by such physical techniques as the application of high gravitational or electrical fields, we are at liberty to announce a 'pure' protein, lipid, or whatever. But we may be sure that this purity is one that we have imposed by our own operational demands on the living material, and that in the cell the interconnections between the macromolecules we have arduously separated are quite as important as the composition of our pure sample. But, especially in the first stages of biochemistry, there was no choice. Before asking *how* the cell functions, we must first know of *what* it is made.

And so we purify the macromolecules, analyse them by breaking them down into their constituent parts, study their structure, and plot the sequence in which the units are joined together. The job of biochemistry as analysis stops there. When, finally, we can put the molecule back into the cell from which it came and try to describe how and why it behaves as it does in any particular biological situation, then we have entered another phase of research.

PURIFICATION AND STRUCTURE OF MACROMOLECULES

Methods of separation may exploit the differential solubility of the macromolecules in different solutions, for instance, salt solutions of different strengths, or in organic solvents (alcohol or acetone); the different stabilities of proteins to acids or alkalis; the different mobilities of the proteins under gravitational and electrical fields (centrifugation and electrophoresis); and the ability of certain substances to absorb some proteins but not others (chromatography). Some of these methods are worth describing in a little detail.

Certain artificial resins, and some materials such as cellulose and aluminium hydroxide, tend, when stirred with a solution containing charged ions, to draw the ions out of solution and bind them to themselves. If the resin is packed into a glass cylinder (column) and the solution poured in on top of the resin column, the liquid that emerges at the bottom is practically free of solute. Such a principle lies behind several domestic water-softening devices. If a protein solution is used, the proteins become bound on the column. They can, however, be washed out again by increasing quantities of water or salt solution. This is the principle of *chromatography*. Each protein passing down the column runs through at a characteristic speed, depending on the concentration and pH of the solution being used to wash it out. Thus the different proteins in the original mixture arrive at the bottom of the column one after the other, and, if the liquid emerging from the column is collected in separate tubes, a few drops at a time, some tubes will contain one of the proteins, some another, and a separation will have been obtained. This elegant method was first used by the Russian botanist Tswett for separating the differently coloured materials present in plant pigments.

Affinity chromatography takes advantage of the fact that many proteins tend to attach themselves preferentially to (that is, have affinity for) particular chemical groups. For example, concanavalin A, a plant protein, can be purified by passing a crude extract containing it through a column packed with resin to which glucose molecules have been bound. The concanavalin molecules will tend to stick to (technically, *bind*) the glucose, whereas most other proteins in the mixture will not and will therefore pass through the column. The bound concanavalin can then be detached and driven out of the column by pouring a strong glucose solution through it. The same principle enables many different types of protein to be simply purified provided that molecules to which the proteins bind can be obtained. As we will see later it is now possible to prepare specific antibodies (themselves protein molecules) to many proteins, and when such an antibody is fixed to a column it can also be used to bind and hence extract a single protein from a complex mixture.

Gel filtration enables one to separate the molecules from one another and at the same time make an estimate of their molecular weight. It involves washing a solution of the macromolecules through a glass column packed with a gel-like substance composed of small beads of porous material. Large molecules pass through the column easily because they cannot enter into the beads. Smaller molecules, however, can penetrate the gel pores, and therefore take a more circuitous and longer route through the column. If a solution of a substance of known molecular weight is also passed through the column, an estimate of the molecular weight of the separated macromolecules can be obtained by comparing the rate at which the substances pass through and are washed out.

Both separation and molecular weight determination can also be achieved by the process known as *gel electrophoresis*, if it is carried out in the presence of a detergent, SDS (sodium dodecyl sulphate). A drop of the solution of the proteins, dissolved in SDS, is placed on to a sheet of a gel made of starch or acrylamide (there is nothing mysterious about such gels; common kitchen gelling agents would do as well – perhaps even commercial jellies! All that they do is provide an inert supporting sheet through which the proteins can travel. How and why polysaccharides tend to form gels is discussed in the next section). The SDS binds to the macromolecules, and coats them evenly with negative charges. If an electric current is then applied to the gel the proteins will migrate towards the positive electrode (anode). Their rate of progress will depend on size and can be compared to the electrophoretic mobility of a substance of known molecular weight (Figure 1). Having separated the proteins by migration in this way in a single dimension the gel sheet can be turned at right angles and a different combination of current and conditions used to drive the proteins in the second dimension. Such *two-dimensional electrophoresis* has a very high resolving power, and more than a thousand different proteins can be distinguished from a crude extract in this way.

The next step after separating a macromolecule and obtaining a 'pure' sample is to try to determine its shape, an important factor in any consideration of the biological function of the macromolcule. Especially with polysaccharides this can be done by investigating their behaviour in solution. Solutions of different-shaped molecules differ in properties such as viscosity, which depend upon the particular shape of the molecule concerned.

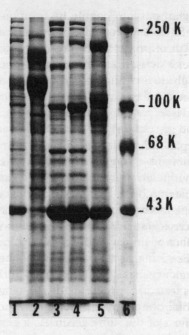

Figure 1. Proteins separated on a gel. Each of the first 5 'ladders' separates a different starting group of proteins. The sixth contains molecular weight markers. K = 1000 and refers to molecular weight.

X-ray diffraction is by far the most accurate of the methods for determining shape, having the ability to discriminate between individual atoms in a large molecule. A sample of the solid substance is bombarded with X-rays which will be scattered as they hit the atoms. If these diffracted rays are directed on to a photographic plate, a pattern can be developed which corresponds directly to the orientation and distribution of atoms within the molecule. We shall see this technique 'in action' later in this chapter. Finally, *electron microscopy* can also be highly informative about the structures of biological molecules and even the position of atoms within them.

The final piece of information needed to determine structure can only be obtained by breaking the molecule down into its constituent parts to enable its components to be identified and to plot their sequence and the nature of their linkages.

A number of properties, common to all types of macromolecule, can be derived from such analyses. All seem to have a certain individuality and many have the ability to specifically recognize and interact with other substances. They are all, to a greater or lesser extent, flexible molecules, capable of adapting their shape in response to their particular microenvironment within the cell. They all have a pronounced intolerance to extreme conditions, a fragility that makes them fall apart if treated harshly. Finally they all may be described in terms of a structural hierarchy of primary, secondary and tertiary structures that we need to look at in a little more detail.

The primary structure of a macromolecule refers to the large number of similar building blocks, each covalently bonded together, giving a stable, firm backbone to the whole molecule. Secondary structure describes how this backbone is folded or pleated in a regular way to allow a large number of weak internal bonds to be formed between different chains or different parts of the same chain. As a general rule, secondary structure is determined by a macromolecular chain folding in such a way as to maximize the number of hydrogen bonds (H-bonds) it can make. (This type of bonding takes place whenever two electronegative atoms, usually O and N, come close enough together to interact with hydrogen and share a positive charge.) There are a variety of other weak bonds which help shape the macromolecular structure, but it is not necessary to discuss them all in detail here.

Tertiary structure describes the way in which these H-bonded chains twist in on themselves to give a complex 3-D shape. A variety of weak bonds and ionic interactions between different parts of the long molecular chain hold it in a tight and well-defined structure, with parts of the molecule buried deep inside, inaccessible to water or other small molecules, while other regions form the surface, where they may interact with all the other ions and molecules which swarm in the microenvironment of the macromolecular chain. Now although the combined strength of many of these weak bonds is quite great, each individual bond requires little energy to be broken or remade and it is this fact that gives the molecules their flexibility. Breaking a few weak bonds will slightly change the shape of the molecule, without significantly altering its overall properties. The importance of this will become clear later. This kind of conformational change may also provide the cell with a method for getting rid of unwanted molecules by exposing previously protected inner areas to degradative attack.

Paradoxically, although these weak bonds are a source of strength and are an essential aspect of the molecule's structure, they are also responsible for the fragility of the molecules, because the bonds depend for their existence on very precise conditions of ionic strength, pH, temperature and the extent to which water is present within their localized area in the cell. A change in any one of these factors may cause destruction of the bonds and an unfolding of the molecule into a naked single chain. This process is called *denaturation* and is generally irreversible.

With these general comments, we can now turn to a consideration of the particular classes of macromolecules in more detail.

CARBOHYDRATES (POLYSACCHARIDES)

We saw earlier that the hexoses, glucose and fructose, could combine by means of a glycosidic linkage to form molecules each containing two sugar units, *disaccharides*. The glycosidic link is formed through the carbon atom 1 of one sugar and the carbon atoms 2, 3, 4, or 6 of a second. The formation of a disaccharide leaves free carbon atom 1 of the second sugar, thus:

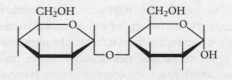

and a further glycosidic link becomes possible:

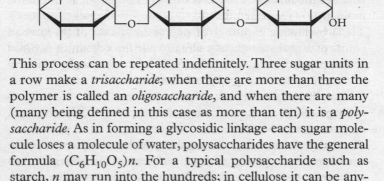

This process can be repeated indefinitely. Three sugar units in a row make a *trisaccharide*; when there are more than three the polymer is called an *oligosaccharide*, and when there are many (many being defined in this case as more than ten) it is a *polysaccharide*. As in forming a glycosidic linkage each sugar molecule loses a molecule of water, polysaccharides have the general formula $(C_6H_{10}O_5)n$. For a typical polysaccharide such as starch, n may run into the hundreds; in cellulose it can be anything from 300 to 2500.

The polysaccharides are widely distributed in plants an animals, both as structural substances (e.g. cellulose), and as food storage compounds (starch, glycogen). As the number of units in the chain increases, the polysaccharide changes its physical properties. Mono- and disaccharides such as glucose and sucrose are water-soluble and taste sweet; as the molecular weight increases, solubility decreases and the taste alters; a 'starchy' taste is quite different from sugar. After a certain point, however, it no longer seems to make much difference how many units more are in the chain. The addition or subtraction of 10, 20, or even 100 units seems scarcely to alter the properties of the molecule. Nor does it ever become possible to refer to 'the' molecular weight of a polysaccharide; a handful of starch will be made up of many chains of varying lengths. The best that can be given is an average figure, an even this tends to differ depending on the source of the sample and the method of purification.

It is clear that, within the polysaccharide chain, there is considerable scope for variation. Thus, as well as the number of units within the chain, it is possible also to alter:

(1) any or all of the individual sugar units (galactose for glucose, for example);

(2) any or all of the points of attachment between the sugars (carbon 1 to carbon 6 instead of to carbon 4, for example);

(3) the geometry of these attachment points (for example, whether the linkage is in the α or β configuration, as discussed previously – see page 30);

(4) by linking carbon 1 of one sugar to one of the carbon atoms of a glucose which is already part of a chain, it is possible to produce *branched chains*. Such branches can be one unit or several units long, and the branches can themselves divide further, like those of a tree (Figure 2a; page 52).

Depending on how these links are made, the resultant macromolecule will differ considerably in properties. Obviously, the nature of the sugar units (and whether they have amino groups attached to them) will affect the chemical behaviour of the macromolecule. But in addition whether the glycosidic bond is α or β, for example, will affect the flexibility of the molecule and whether individual molecules can lie close together in a rigid configuration.

So how does one set about discovering the structure of a molecule in which so many permutations are possible? The temptation to despair when faced with a molecule which can be composed of 1000 units, any one of which may be one of sixteen different hexose isomers (to say nothing of the pentoses) linked in any one of five possible ways, each with an isomeric twin, to its neighbour, must be resisted. And indeed, it turns out that the problem is much harder than for proteins which may be considerably larger molecules. The only saving grace is that despite the fact that the number of possible permutations available is rather greater than those one is faced with on the average lottery ticket, only a relatively small number actually seem to occur, and even amongst these few naturally occurring combinations, minor variations do not seem to be very significant for the state of the final product. But one may well blench at the problem of actually determining the structure of these mammoth molecules. Fortunately, though, as we have seen, quite elegant methods do exist.

The first step is obviously to determine which of the

monosaccharide sugars are actually present, and this can be done easily enough by hydrolysing the polysaccharide (quite strong hydrochloric or sulphuric acid is often necessary). On hydrolysis, all the glycosidic bonds between the sugar units break, releasing the constituent sugars, which can then be separated and identified. If only one sugar is present (when the polysaccharide is called a *homopolysaccharide*) then one is nearly home and dry. If two or more sugars are present, it becomes necessary to determine the details of the order in which they are joined together. Fortunately, it is usually the case that the various sugars are linked to form a regular, simple, repeating sequence; for example, an alternating series of glucose and galactose units. Sequences are studied by a modification of the hydrolysis procedure, in which milder conditions and shorter hydrolysis times are used. Under such circumstances the polysaccharide chain is not entirely broken down to monosaccharides, but is instead only partially hydrolysed into many small fragments of chain, each containing several sugar units. Just as with the total hydrolysate, the different fragments of the broken chain can be separated and the structure of each determined.

To take a hypothetical case, suppose the original polysaccharide contained glucose and galactose in equal proportions. We might then find amongst the chain fragments the oligosaccharides

gluc-gluc,
gluc-gluc-gluc,
gluc-gluc-gluc-gluc, and,
gluc-gluc-gluc-gluc-gal,

and also,

gal-gal
gal-gal-gal, and,
gluc-gal-gal.

We might then suspect that the original polysaccharide was composed of separate glucose and galactose chains, with a gluc-gal branching point where one chain joined the other. Alternatively, if the products of hydrolysis were the fragments

gluc-gal-gluc,
gal-gluc-gal,
gal-gluc-gal-gluc, and so forth,

one would clearly be dealing with a polysaccharide in which each chain contained an alternating sequence of glucose and galactose residues.

In more sophisticated versions of this 'jigsaw' technique, the biochemist seeks for specific enzymes which will hydrolyse bonds between some sugar residues but not others. In all such methods, however, one ultimately has to solve the intricate puzzle of fitting the bits together again in what seems the most likely manner – a task which, these days, computers make rather easier.

To complete our picture of the polysaccharide, a figure for the molecular weight is needed. Some information is provided by the chemical evidence, if we know chain lengths and numbers of branching points, but physical methods of the sort discussed in the previous section are normally indispensable.

The use of such techniques has provided the basis for our understanding of the composition of polysaccharides. New and hitherto unknown polysaccharides are constantly being found as more and more plant, bacterial, and animal tissues are analysed, and many of these have highly complex patterns of constituent sugars and branching chains. But the most frequently occurring of the polysaccharides are rather simpler in structure. There are three: two of plant origin, *cellulose* and *starch*; and one from animals, *glycogen*.

Cellulose

Think of a naturally occurring fibre, and it is odds-on that it will be made of cellulose. Cotton, flax, wood, paper, even most industrially-made fibres, are all cellulose; the hairs of the cotton seed plant, in particular, are over ninety per cent pure cellulose. There is more of it in the world than any other organic chemical, for it forms the structural framework on which plants (and some bacteria) are constructed.

Hydrolysis of cellulose yields nothing but glucose. Partial hydrolysis gives glucose 1–4 β-linked di- and trisaccharides. So cellulose has a primary structure composed of straight chains of 1–4 β-linked glucose units; molecular weight estimations, which give a value of 50 000 to 400 000, imply that there are between 300 and 2400 glucose residues per molecule.

The value of cellulose as a natural fibre lies in just this fact of long, straight glucose chains. It is the β-linkage that gives the cellulose chain its ability to lie like a flat ribbon, enabling molecules to be packed together, all lying in the same direction, to form a crystal-like thread with a strength somewhat greater than a thread of high-grade steel of the same diameter. The long threads are composed of oriented bundles of cellulose molecules, each individual bundle being known as a *micelle*. There are also H-bonds formed between the glucose units of the individual molecules within the micelle, thus increasing its rigidity. Individual micelles can easily be seen when electron microscope pictures of cellulose fibres are made, and crystallographers, by measuring the dimensions of the micelles, have been able to deduce a great deal about the intramolecular configuration of cellulose fibres. From the point of view of the biochemist, interested in natural structures, it is the rigidity of the fibres which is of most interest, because it is this that enables cellulose to provide the framework for the plant cell. When the uniform orientation of the molecules is lost, and the micelles are shuffled so as to point in many different directions, the fibre loses its strength; the product of this disorientation has commercial value as cellophane (Figure 2c).

Apart from cellulose, many other polysaccharides help maintain plant structure. Among these are *arabans* and *xylans*, which give wood many of its typical properties and are made from the pentose sugars arabinose and xylose.

Chitin serves the same function in an insect exoskeleton as cellulose does in the plant. Chitin differs only in that one of the ring hydroxyls in each of the repeating glucose units is replaced by the acetamido $NHCOCH_2$ group, but this small difference means that more H and other bonds can now hold the chains together, making for even greater rigidity. The acetamido group also bonds very easily to proteins and in fact chitin is usually found linked to protein, to form a *proteoglycan*.

Starch

Starch typifies the role of the polysaccharides as food stores in plants; storage organs such as tubers, fruits, and seeds may contain up to seventy per cent of starch, which can be broken down into glucose and used for food when the need arises. It is easy to see how this storage role is aided by the use of a molecule which is compact, insoluble but readily degradable. Unlike cellulose, starch is a granular material with no trace of organized crystalline structure, yet it too is composed of chains of glucose units, 1–4 linked. The critical difference lies in the fact that the chains in starch are α-linked. Starch is a mixture of two polysaccharides, the one a straight chain called amylose, about 250–300 glucose units long, with the chain folded and packed down into a helical shape. The other is a branched chain called amylopectin in which straight 1–4 linked chains are joined together by 1–6 linked branching-points, one every twenty-five glucose units or so, to produce a molecule which when drawn looks more like a gorse bush than anything else (Figure 2d). It is these branching points which make the molecule easy to degrade, as they provide sites for the beginning of breakdown, as we shall see later. About seventy to eighty per cent of starch, by weight, is made up of the branched chain amylopectin. The two chains can be separated simply by adding hot organic solvent to a starch dispersion, when the amylose is precipitated. The ability of starch to form pastes depends on the highly branched amylopectin.

While starch is the commonest storage polysaccharide, fructose polymers are also sometimes found, for example in

Figure 2. Carbohydrate molecules

(a) Sugar units can be arranged in several ways:

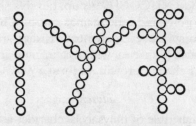

(b) In cellulose the long, straight molecules lie side by side:

(c) While in cellophane they are disorganised:

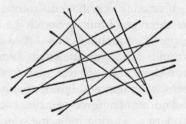

(d) In amylopectin the molecule is highly branched:

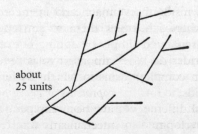

about
25 units

Jerusalem artichokes and asparagus, possibly accounting for the rather distinctive taste of these vegetables.

Glycogen

Animals store polysaccharide mainly in the liver and muscle, as glycogen. This is a glucose polymer similar to amylopectin, but even more highly branched, with a dividing point every 12–18 units along the chain.

Other polysaccharides

There is another type of polysaccharide that has a supportive or protective role in plants and animals, although by a totally different method to cellulose. These are the gelling polysaccharides of which pectin (generally known for its use in jam-making) is an example. It is found in and between the cell walls of higher plants and is composed of chains of galactose, arabinose, and the sugar acid galacturonic acid. The glucosaminoglycans are found in the mucous linings of the respiratory and digestive tracts of animals. Hyaluronic acid and chondroitin sulphate provide the lubrication in knee joints and the agars prevent seaweeds drying out at low tide. These gelling polysaccharides differ from the fibrous molecules in that the covalently bonded portions that make up their primary structures are connected by areas where there is a three-dimensional network of weakly bonded units, often forming helical shapes. This 3-D network can trap large quantities of water, forming a gel, and this confers elasticity on the material, enabling it to stretch and then bounce back to its original shape. This is of obvious importance where a protective coating must not restrict movement, such as inside the stomach and in the cell wall of red blood cells, where it enables them to squeeze through the narrow capillaries.

Polysaccharides do have important roles at the surfaces of animal cells, in complex forms in which they are attached to proteins or lipids, to form *glycoproteins* or *glycolipids*; and where their structural differences have been exploited during evolution in the development of mechanisms whereby cells, when they come into contact one with another, 'recognize' each other so as to distinguish like from unlike – an important property in relationship to growth and development and also for the body's immune system. However, in general polysaccharides play a less important structural role in animals than in plants. This role is filled in animals by lipids and proteins, which offer greater potentialities for physical and chemical modification than the carbohydrate polymers, and hence enable the animal to achieve a more flexible and adaptable existence. It is perhaps significant that the only animals to use polysaccharides as structural components are those, like insects and arthropods (crabs, for instance), where the tough, brittle chitin is used as a so-called external skeleton. It is just this existence of the exo-skeleton which has placed such sharp limitations on insect size. For animals at least, there is no future to being built on a poly-saccharide framework.

PROTEINS

The name protein, coined in 1838, comes from the Greek for 'first things', and it was a fortunate choice. Most of life revolves around the activities of the proteins; the energy of the cell goes largely first into making them, and then into using them to per-form a multitude of diverse jobs. So much so, indeed, that a measure of the importance of the proteins is that we should be hard put to it to decide whether they evolved to suit the de-mands of living, or whether life developed to fit the require-ments of the proteins. (Of course, that is a *biochemist's* view – molecular biologists differ from biochemists in this respect; they would give primacy to the nucleic acids, for reasons that will become apparent as this book proceeds.)

What are proteins? Macromolecules which on hydrolysis yield practically nothing except a mixture of amino acids. In the protein, the amino acids are linked by *peptide bonds*. The peptide bond is formed by joining the amino group of one amino acid to the carboxyl of a second. Here is the equation of peptide bond formation between glycine and alanine.

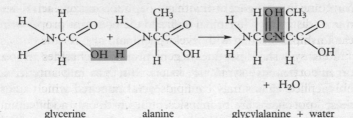

+ H_2O

glycerine   alanine   glycylalanine + water

The peptide bond, then, is the linkage

$$-\underset{\underset{O}{\|}}{C}-\underset{\underset{H}{|}}{N}-$$

between two amino acids, and the resulting molecule is called a *peptide*. The linkage is covalent, and the peptide bond fills the same function for proteins as does the glycosidic bond for polysaccharides. For each peptide bond formed, one molecule of water is released.

As two amino acids are present in the molecule shown in the equation, it is specified more precisely as a *dipeptide*, in this case glycylalanyl dipeptide. The peptide link between the two amino acids still leaves free the $-NH_2$ amino group of the glycine, and the $-COOH$ carboxyl group of the alanine. Both of these are thus free to participate in further peptide bond formation with other amino acids. For example, with serine, either of the two reactions

glycylalanine + serine $\longrightarrow$ glycylalanylserine + H_2O (2)

serine + glycylalanine $\longrightarrow$ serylglycylalanine + H_2O (3)

is possible. The two tripeptides, though both containing the same amino acids, are different compounds and will have different physical and chemical properties. For a tripeptide, then, six different isomers are possible, the two shown and also alanylserylglycine, alanylglycylserine, serylalanylglycine, and glycylserylalanine. (We can in future abbreviate these cumbersome names by using the abbreviations, or letter symbols, shown in the list of amino acids, Table 3, page 34.) From these formulae, it is apparent that each peptide contains an N-terminal end and a C-terminal end, and the convention is to write the formulae starting with the N-terminal.

Relatively short peptides can be prepared by rather tedious organic chemical synthetic routes. But naturally occurring polypeptide chains may contain several hundred amino acids; these long chains are proteins. Peptides with only a few amino acids (generally two or three) do occur in the cell, and there are a number of important peptide hormones with as many as fifteen. But the biochemist's major concern is with the proteins.

We saw that for a tripeptide, six isomeric structures were possible. As the number of amino acids increases, the number of possible isomers grows enormously. For a relatively modest protein – with a molecular weight of 34 000, and with 288 amino acids, but made up of only 12 different amino acids out of the possible 20 – the number of different isomers is 10^{300}. If only one molecule of each isomer were to exist, the total mass would be some 10^{280} grams. As the weight of the earth is only 10^{27} grams, it is very clear that only a tiny fraction of these isomers in fact exist. Even so, the number of proteins in existence is a formidable one. Several thousands have been more or less purified, and it can be calculated on the basis of genetic data that the total number of chemically and physically distinguishable proteins present in the cells of the human body is of the order of 80–100 thousand, whilst there is also a good deal of evidence to show that even apparently similar proteins from different species are not quite identical in structure and amino acid composition. So that, within the compass of life on earth, there are certain to be many millions of individual and unique proteins.

But here we run into one of the prime difficulties of protein biochemistry – deciding precisely what we mean by an 'individual and unique protein structure'. Granted that every polypeptide chain with a defined amino acid sequence is a single protein, and that any chain which differs from it by so much as a single amino acid altered or out of place is a different one (and we mentioned earlier the evidence of sickle-cell anaemia to show that this can in fact be the case), how do we distinguish between the chains; how do we obtain a pure protein?

The analytical techniques that we have discussed, of gel filtration, chromatography and electrophoresis, are all important here, where it is vital to use methods which do not destroy the delicate protein chains in the process of handling them. Nowadays, the battery of available techniques makes the purification of most soluble proteins a tolerably straightforward affair. The first necessity is to have some means of recognizing the specific protein one is trying to purify. This is easy if the protein is an enzyme, because then it, and no other protein, will catalyse a specific reaction, and the presence of the enzyme in a mixture of other proteins can thus always be noticed. Any protein fractions which do not contain the ability to catalyse the reaction do not contain the enzyme. Similarly the protein may be tagged if it is one of those containing heavy-metal ions, such as iron, which are firmly bound to it (*metalloproteins*); other such tagging groups are the phosphates of *phosphoproteins* (which are bound to serine and threonine residues in the amino acid chain) and the carbohydrates of *glycoproteins*. As we will see, these 'extra' components are of enormous biological significance for the functioning of proteins in their natural environment of the cell.

Having obtained the protein, how do we know if it is pure? A lot depends on exactly what we mean by pure in this context. In the early days of protein chemistry much attention was paid to ridding the protein of all traces of non-protein materials, such as metal ions or fragments of lipid or carbohydrate. But it is now clear that in the living cell itself these components often occur bound into extremely close association with the protein. To remove them, even when this can be done, may produce a satisfactorily chemically pure amino acid chain, but that is not really what one is after; the important thing is to isolate as close an approximation to the native material as possible – even if this does mean taking it with its warts and all.

One can only tell if the protein is really free of other proteins by trying to go on fractionating it. If it doesn't split into several components under ultracentrifugation or electrophoresis or chromatography, it is probably a single molecular species. Pure proteins can normally be crystallized, but even this is not an absolute indication of purity; there may still be five or ten per cent of another protein present that cannot be removed without great difficulty. But only when one is sure that the protein is pure can any attempt be made to determine its structure.

As with polysaccharides, there are two parts to the structural problem, determination of the primary sequence, the order of amino acids along the protein chain, and the shape of the molecule in space, its secondary and tertiary structures. Many of the special properties of proteins arise from the nature of the peptide bond. This peptide bond limits the ways in which the polypeptide chain can fold. All four atoms in the $-C_\alpha O-NH-$ group lie in the same plane and, as a consequence, the polypeptide chain can fold only about $C_\alpha-N$ and $C_\alpha-C$ bonds. Chemical and X-ray diffraction studies of proteins provide information on how amino acid residues affect protein folding. For example, the large non-polar residue of proline has the property of forcing a bend in the main chain and disrupting the H-bonds between different parts of the chain. The small neutral polar residues of serine and threonine can, by hydrogen bonding to other like amino acids, cross-link them in the same or in two different chains. If in the same chain, this cross-linking may bring very distant regions of the primary amino acid sequence into close conjunction, which helps explain why the function of a protein cannot be understood from its primary sequence alone without taking account of the folding.

Obtaining this information is not easy, and the only saving grace in the matter of amino acid sequence is that polypeptide chains are never branched as polysaccharides are; there is only a linear sequence to cope with. This isn't to say that many proteins do not consist of several polypeptide chains; they do, but, as we shall later see, they are joined by linkages other than peptide bonds; they belong to the secondary structure of the protein.

The first step towards determining the amino acid sequence is to find out which amino acids are present, by hydrolysing the protein and separating and measuring each of the amino acids.

However the information provided by such analysis is limited as it cannot provide information on the sequence.

One must, then, resort to the 'jigsaw' techniques we mentioned in connection with polysaccharide structure, and try to build a picture of the complete protein by combining the information gained from an analysis of many small sections of it. It is of importance here to find which amino acids form the N- and C-terminal ends of the molecule. Fortunately, there exist certain enzymes and chemical reagents which combine specifically with *either* the C- *or* the N-terminal amino acid. Use of these enables one to decide which amino acid begins, and which ends, the protein chain.

Once the terminal amino acids are known, the next step is to make a series of partial hydrolysates, splitting the protein into fragments two to five amino acids long, and determining the sequences in each of these chainlets separately. Partial hydrolysates can be made with acid hydrolysis for short periods of time or at low temperatures, but there also exist several enzymes which break down protein chains, and these enzymes have the virtue of being fairly discriminating about which bonds they will and won't attack. Trypsin, for example, an enzyme present in digestive juices, splits any peptide bond where the amino acid at the C-side of the CO.NH bond is either lysine or arginine, thus producing peptides whose C-terminal amino acid is lysine or arginine. Chymotrypsin, another digestive enzyme, does the same where the amino acids are phenylalanine or tyrosine. Many other protein-splitting enzymes can be used similarly. All that remains, when one has obtained the partial hydrolysates from these sources, is to separate them and determine each individual sequence before trying to put them all together in the most likely structure.

Thus the experimental strategy for determining the amino acid sequence of a protein is to divide and conquer, splitting it into shorter peptides which can then be individually sequenced. Frederick Sanger and his associates at Cambridge completed this little puzzle for the small protein insulin, which has less than sixty amino acids and a molecular weight of only

12 000, in rather less than a decade, and in doing so became in 1956 the first people ever to publish the full sequence for any protein. By the time Sanger was given a Nobel Prize for this achievement in 1959 (this was Sanger's *first* Nobel; later he got a second – an almost unique distinction! – for deciphering nucleic acid sequences), several other protein structures were far advanced. The procedure is now automated, and complete or nearly complete sequences are now known for thousands of proteins in many different species.

Meanwhile, the way towards a quite different approach to protein sequencing has been opened by increasing knowledge of the structure of DNA and RNA. We will see later how it is that DNA and RNA sequences determine protein sequences; for now the point is that one can predict a protein structure on the basis of a nucleic acid structure. In 1978 the first virus nucleic acid sequence was completely decoded, and hence it became possible to know the viral protein sequences too. Peptide sequences once determined are stored in computerized data banks, and as each new sequence is found it can be directly compared with already known ones, thus also helping speed up deciphering the protein. The result is that a procedure that once took a decade and won a Nobel Prize is now a matter of weeks and scarcely rates the award of a PhD.

How about the higher order structure of proteins? It is sometimes claimed that primary sequence determines all higher orders, but this is true only to a limited degree. We have said that secondary structure is determined by simply folding the primary chain so as to achieve the largest number of H-bonds between different parts of the chain. However, in the proteins another much more powerful bond is provided by the amino acid cysteine:

$$HS.CH_2\ \underset{\underset{NH_2}{|}}{CH.COOH}$$

The —SH group in one molecule of cysteine can readily combine with that of a second to produce a 'disulphide bridge':

$$CH_2.CH.COOH$$
$$\mid \qquad \mid$$
$$S \qquad NH_2$$
$$\mid$$
$$S \qquad NH_2$$
$$\mid \qquad \mid$$
$$CH_2.CH.COOH$$

If the two cysteines are in different parts of a protein chain, or are in two separate chains, the chains bend towards each other and are held together by the bridge. Many proteins contain such cysteine bridges which are covalent bonds and cannot easily be broken. Thus in insulin there are two disulphide bridges between the two chains of which the molecule is composed, whilst a third bridge connects two distant portions of one of the chains. In ribonuclease, which is a single-chain molecule, four disulphide bridges twist the chain into a snake-like

Figure 3. Two protein chains

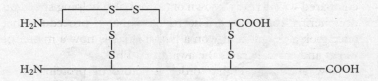

(a) Insulin: two chains bound together by disulphide bridges

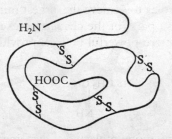

(b) Ribonuclease: one chain twisted into a ball by disulphide bridges

configuration (see figure 3), whilst the enzyme chymotrypsinogen has five bridges. Apart from disulphide bridges the other weak bonds we have mentioned are also present helping to maintain secondary and tertiary structure. These bonds often involve charged groups, and the charge distribution across the protein chain is strongly affected by the microenvironment of the molecule – for instance, the acidity or alkalinity of the solution, the presence of cations like Ca^{2+}, Mg^{2+}, and heavy metals. Thus while it is true that in any given microenvironment the shape of a protein is determined by its primary amino acid sequence, if that environment is altered, the shape changes – a fact that can be of major biological significance. And to date attempts to predict the three dimensional folded structure of proteins from their primary sequence have met with only limited success.

Figure 4.

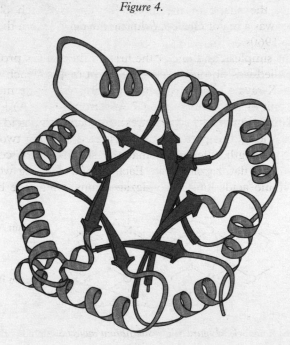

According to their shape, protein molecules are divided into fibrous and globular proteins. As their names imply, the fibrous proteins have straight chains, the globular proteins have chains which are coiled together into irregular ball-shaped molecules. The skeleton drawing of Figure 3 is thus misleading, and proteins are better envisaged as in the space-filling or ribbon-like diagrams of Figure 4.

Amongst the fibrous group are the structural proteins of the body: keratin in hair and nails and wool, collagen in skin and connective tissue, elastin in tendons and arteries, fibroin in silk. These proteins are all insoluble, relatively tough molecules resistant to acids, alkalis, and quite high temperatures. The resistivity of these molecules, like that of cellulose, is a reflection of their ability to form ordered, semi-crystalline structures. The advantage to the biochemist is that it is possible to make X-ray photographs of these crystal-like formations and use them to deduce the shape of the molecules, and research on these shapes was a major challenge during the period from the 1930s to the 1960s.

The simplest, and one of the first, of the fibrous proteins to be studied was fibroin (see Figure 5), from silk, which showed under X-rays a regular repeating pattern within the molecule, each unit of the pattern being 7 Ångstrom units (Å) long (an Ångstrom unit is 10^{-8} centimetres). As each amino acid residue is 3.5 Å long, this pattern was interpreted as being two amino acids in length, suggesting that the silk molecule could be drawn as a flat, zigzag chain. Each complete zigzag would be two amino acids long. Two zigzag chains, lying side by side,

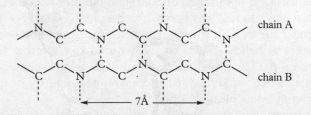

Figure 5. The silk fibroin molecule

would stick together at alternate 'zags' by reason of hydrogen bonds. Such a repeating pattern could result in an extended meshwork of chains, and a description of this sort is sufficient to explain the X-ray diffraction patterns made by fibroin.

Other molecules, though, are more complex. The classical experiments were made with keratin. The keratin fibre can, when soaked in water, stretch to as much as twice its original length. The unstretched form is called α-, the stretched form, β-keratin. Astbury, at Leeds, who made these studies, found that the stretched β-keratin gave an X-ray picture practically identical to that of silk fibroin. The α-keratin, then, must be present in some more contracted form than that of the silk fibroin. Great efforts were made to interpret the structure and many ingenious suggestions put forward until, finally, in 1951 Pauling and Corey in America concluded that the only pattern which met all the requirements was if the molecule was coiled, like a loose spring, into a spiral conformation which they termed the α-helix. The helix can be shown diagrammatically as in Figure 6(a) or in more detail as in Figure 6(b). In its natural shape the α-helix makes five complete turns for every 18 amino acids, or one turn for 3.7 acids, and the amino acids which fall immediately above and below each other as the helix ascends clutch hold of one another by means of hydrogen bonds, and so keep the whole structure stable. When the α-keratin molecule is stretched, the helix is straightened out like a pulled string, until ultimately it assumes the 'unwound' fibroin configuration of β-keratin. Pauling and Corey later refined their picture of the helical protein molecule by showing that the helix itself was not formed on a straight but on a helical axis, so that it too gradually twisted, like the coiled-coil of a lamp filament (figure 6c). They postulated that it would be possible for several helical chains to twist round one another like a woven rope. The protein in hair and nails, they suggested, was formed when six keratin α-helices wrapped themselves round a seventh into a pleated bundle.

It is hard to overestimate the importance of the discovery of these helical structures. As more and more macromolecules

Figure 6

(a) Helical structure of proteins. The helix turns five times for every 18 amino acids, and is held in place by hydrogen bonds. (b) The Pauling α-helix: a close-up view. (c) Some super-helices.

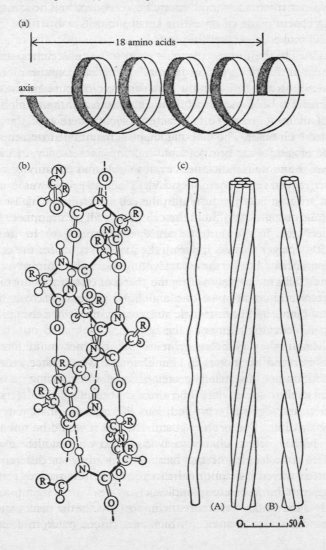

came to be studied it became clear that the helix is a recurring motif within biology, culminating, as we will see shortly, in the most famous of all – the DNA double helix. But there is nothing mysterious about its appearance, because it turns out that simple mathematical and energetic considerations mean that long chains made of repeating small subunits will always tend to fall into a coiled, helical pattern.

Thus all the fibrous proteins seem to be built on a pattern similar to that of keratin or fibroin. The helical configuration of α-keratin is also found in the globular proteins, but here the entire helix is bent and bunched together into a compact globule as if the taut rope of the keratin molecule were being neatly coiled by a sailor. The globular shape is maintained by a multitude of disulphide bridges and weak ionic interactions. In this form, the protein molecule is both soluble and chemically very reactive. Enzymes, hormones, and, in fact, all protein molecules that are normally found within the cell are globular, whilst the fibrous proteins help build the cell wall or fill up the spaces between cells. In the same Cambridge laboratory where, in the 1950s, Sanger was deciphering the insulin sequence, the crystallographer Max Perutz was devoting the major portion of his researching life to discovering the shape of one of the most important and commonest of the globular proteins, the one that gives blood its characteristic red colour, the iron-containing, oxygen-carrying pigment haemoglobin.

Many other globular proteins will be met in action as enzymes and hormones in later chapters. When these protein functions are discussed it will become clear that it is not sufficient to speak only of primary, secondary and tertiary structures, together with additions like metals, phosphate or carbohydrate. The proteins themselves are often combined in yet higher order (*quaternary*) complexes, in which many different molecules of the same – or sometimes different – proteins, given the right microenvironment, 'self-assemble' into large multiprotein complexes such as enzymes, membranes and subcellular structures. Figure 7 shows one such. Ribosomes, for instance, which are discussed in the next

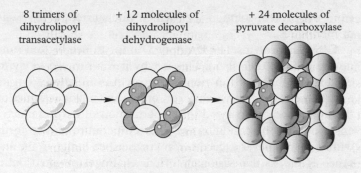

8 trimers of dihydrolipoyl transacetylase

+ 12 molecules of dihydrolipoyl dehydrogenase

+ 24 molecules of pyruvate decarboxylase

Figure 7. The pyruvate dehydrogenase complex

chapter, are complexes of protein with RNA in which as many as thirty different protein molecules wrap themselves together into a precise, and seemingly unique configuration. In these higher order complexes the world of chemistry and linear molecular sequences is increasingly abandoned in favour of the world of architecture, in which relationships of the parts of molecules to one another in three-dimensional space becomes all-important.

It is worth emphasizing here, though, the particular sensitivity of the tertiary and quaternary structures of these proteins to factors which affect the weak bonding which holds them together, especially changes in pH and temperature, which rapidly denature most globular proteins and multiprotein assemblies. A simple example is the 'curdling' of milk when vinegar is added – the milk proteins clot together and separate from the whey. A similar process occurs when protein solutions are heated; the weak linkages begin to break at temperatures much above 60–70° C, and again the proteins tumble out of solution and lose their structure. The characteristic appearance of scrambled egg is due to the flocculent precipitate of egg proteins that occurs on heating. Eggs also provide another example of the destruction of higher order protein structure, produced by as simple an activity as steady shaking of the solution – when this is done by beating egg white, protein begins to precipitate, and the resultant material, quite different from the protein solution with which one started, is the basis of meringues.

Globular proteins, like Humpty Dumpty, once broken cannot be put together again. This is why it is a mistake to try to discuss the shape or function of a protein in isolation from its environment. And, as we shall see, one of the major activities of the living body is to avoid the denaturation of its proteins by ensuring that the conditions of pH, temperature and so forth within the cell never exceed certain prescribed limits. In the absence of life, as that visionary Friedrich Engels pointed out a century ago, the proteins are inherently unstable.

NUCLEIC ACIDS

The next class of giant molecules we have to consider is that of the nucleic acids. (It is because this book is about biochemistry rather than molecular biology that these particular actors on the stage of life have put in so late an appearance!) In 1868 Friedrich Miescher in Tübingen isolated from the nuclei of pus cells collected from discarded surgical bandages an unusual phosphorus-containing compound which he called nuclein. He went on to show that a similar substance was present in many other less distasteful materials, notably salmon sperm, and it became apparent that *nucleic acids*, as the new substances became known, were in fact widely distributed in living tissues. Although they were later found to be not entirely confined to the cell nucleus, the name remained as a general one for this entire group. When, much later, the role of the nucleic acids in heredity and in protein synthesis came to be understood, the name acquired, indeed, a new significance and justification.

There are two different classes of nucleic acid, DNA (deoxyribonucleic acid), consisting of the purine bases adenine and guanine and the pyrimidine bases cytosine and thymine, together with the sugar deoxyribose and phosphoric acid. The second, RNA (ribonucleic acid), contains uracil instead of thymine and ribose instead of deoxyribose.

DNA and RNA occur in the tissues in combination with varying amounts of protein, and such protein, especially that found attached to DNA in the cell nucleus, has quite characteristic properties; notably, it is very alkaline, containing large amounts of the amino acid arginine. Such proteins are known as *histones*. In extracting the nucleic acid from the tissue, one of the problems is to remove the tightly bound protein without using methods so drastic as to harm the nucleic acids themselves. This problem is made easier by the fact that nucleic acids are considerably more robust than proteins; they are stable to mild acids and alkalis, and to heating to almost 100°C, procedures which rapidly wreck the delicate protein molecules. Thus RNA may be extracted from yeast by dissolving in a detergent solution at 90°C, whilst DNA can be prepared from thymus by extraction with alkali or salt solution followed by chloroform.

The material that is obtained by these methods is a whitish, fibrous substance (DNA rather resembles asbestos in appearance), quite stable, which can be stored for longish periods without coming to much harm (which is why the plot of *Jurassic Park* was even remotely credible!). Its molecules have a weight in the order of tens of millions and, like proteins or polysaccharides, have a primary structure consisting of a repeating series of simple units joined into a chain. Although, as we shall show, many different DNAs and RNAs exist, the problem of isolating one DNA from a mixture of others, so critical in protein biochemistry, scarcely exists. This is at least partly because the physical properties of the different molecules do not differ sharply enough from one another to make it possible to exploit differences in solubility or stability as can be done with the proteins. Chromatography of DNA does reveal that the 'total' DNA obtained from any tissue can be split into a number of slightly differing fractions, but it has become customary to refer to the 'total' DNA of any organism almost as if it were a single molecular species, or at best two. As will become clear in the next chapter, virtually all of a plant or animal cell's DNA is concentrated in the nucleus, but a small quantity is found outside in cell organelles called mitochondria (and in plants, also chloroplasts) and the nuclear DNA of any cell differs somewhat from its mitochondrial DNA.

Proteins consist only of amino acids, polysaccharides of sugars, but nucleic acids contain purines, pyrimidines, sugars, and phosphoric acid. Thus the basic pattern of the repeating unit is less simple than that of other macromolecules. It was early apparent that the base, sugar, and phosphate were combined as nucleotides, where the purine or pyrimidine is linked through the sugar to the phosphate (see page 37). How are the nucleotides connected to each other, though?

It was clear by the 1940s that some relationship between the four bases must exist, because it was noticed by Erwin Chargaff in New York that for RNA the amount of (adenine plus cytosine) always equals (guanine plus uracil), whilst for DNA an even more precise relationship holds, so that the amount of purines and pyrimidines present in any sample is always identical and the ratio of adenine to thymine and of guanine to cytosine present is exactly one. Also, DNAs tend to fall into one of two main groups – either one in which there is more adenine and thymine than guanine and cytosine, or the rarer one in which guanine and cytosine are the two main components.

For RNA, each nucleotide can be drawn like this:

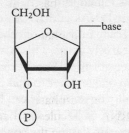

where carbon atom 1' of the sugar is attached to the base, and carbon atom 3' to the phosphate. When the nucleotides join to form RNA, the link is between the phosphate of one nucleotide and carbon atom 5' of the second sugar:

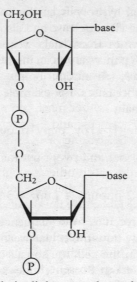

Thus the primary chain link runs through the sugar phosphates, and can be shown schematically:

sugar-base

/

phosphate

/

sugar-base

/

phosphate

/

sugar-base

/

phosphate

The variable in RNA molecules is then the order in which the four bases are arranged along the chain. Although there are only four bases, the number of available permutations down a chain of several hundred is obviously very large, though smaller than for proteins. Sequencing depends on the types of analyses we are now quite familiar with. Hydrolysis, which splits the RNA into its residual nucleotides, gives a modicum of information; partial hydrolysis, in this case using a variety of enzymes, including ribonuclease and also some potent nucleic acid-splitting enzymes from snake venom, is of more value. Ribonuclease from pancreas has an interesting specificity, for it seems preferentially to attack bonds between pyrimidine nucleotides and other bases. For example, the enzyme will split the nucleic acid chain

$$\text{pupupy} \mid \text{py} \mid \text{py} \mid \text{pupy} \mid \text{pupupy} \mid \text{py} \mid$$

(writing pu for purines and py for pyrimidines), along the lines drawn, to give

$$2 \text{ pupupy} + \text{pupy} + 3 \text{ py}.$$

The first RNA molecules to be sequenced in this way were a group of relatively low-molecular-weight substances which occur in solution in the cell, the so-called transfer RNAs (see Chapter 10) with about 80 nucleotides each, whose structure was worked out at all levels in the 1960s. RNA molecules can now be rapidly and directly sequenced by using the high resolving power of gel electrophoresis. One end of the chain is first labelled with a radioactive group, and the labelled chain is then partially digested at one of its four bases using a specific enzyme, or the chain can be specifically fragmented by chemically modifying one of the four bases. The backbone of the chain is then split at the site of the damaged base. The sets of fragments obtained are then resolved by gel electrophoresis and compared using computerized data bases with already known and sequenced fragments. As will become clear later, because of the relationship between protein and RNA and DNA sequences, once one is known it becomes possible to deduce the other, so protein and nucleic acid sequencing each support the other.

The structure of the repeating unit of DNA is similar to that of RNA, with the simple replacement of ribose by deoxyribose. Thus, like RNA, the DNA chain is linked through a 3'—5' sugar phosphate bond, with the purine and pyrimidine bases tacked on to carbon atom 1' of each deoxyribose residue (see diagram on page 37 and Figure 8). But why, if this is all, the peculiar regular ratios that seem to exist between the amounts of the different bases in DNA? And what are the aspects of the secondary and tertiary structure of DNA, whose fibrous character and behaviour in solution are those of a rigid, linear molecule, like a stiff rod?

The crystalline properties of DNA, in at least some of its forms, make it a good target for examination by X-ray crystallography, and studies made by Rosalind Franklin and by Maurice Wilkins in London in the early 1950s indicated that there was a regular repeating sequence within the molecule every 28Å. Now the distance between two nucleotide residues, from phosphate to phosphate, is only 7Å, so in some way there must be a strong repeating tertiary structure to the DNA. James Watson and Francis Crick in Cambridge used Franklin's results to solve the problem in 1953. The papers in which these findings were reported, in the scientific journal *Nature*, have become some of the most famous in the history of twentieth-century science, for the structure proposed for DNA was immediately seen by Watson and Crick to provide a mechanism for fundamental processes of accurate cell replication during growth and reproduction, as well as, in due course, for protein synthesis.

Franklin died in 1958, and Wilkins, Watson and Crick were given the 1962 Nobel Prize for the discovery of the structure of DNA. The events leading up to the discovery have been much debated by historians and sociologists of science and have been the focus of fierce political discussion, for essentially Watson and Crick were led to their proposed structure on the basis of Franklin's data which they saw and applied without her knowledge. Watson's account of the affair made a popular book – *The Double Helix* – in the 1960s; Crick and Wilkins have also written their accounts. Franklin's case was taken up by her biographer, by a number of feminist historians and sociologists of science (as an exemplar of the exclusion of women from science), and in a popular television docu-drama in the 1980s.

The DNA structure that Watson and Crick proposed, as all the world now knows, was a helix – at almost the same time as

Pauling and Corey found that the α-helix also gave the answer to the keratin problem for proteins. The DNA helix as proposed by Watson and Crick turns once every 10 nucleotides (proteins, it may be remembered, manage to make the turn in 3.7 amino acids), and it contains a double strand of two DNA chains twisted together about the same axis. All the way along, the two chains link together through their bases. When Watson and Crick came to make scale models of the structure they proposed, they found that the two chains could only fit together if every adenine (A) on one chain was opposite a thymine (T) on the other, and every guanine (G) opposite a cytosine (C). Then, and only then, the two strands would latch together, and hydrogen bonds between each pair of bases would hold them firmly in place. By contrast, if one considers adenine and cytosine, which do not form base pairs, the three atoms necessary for H-bond formation cannot be set into a straight line, and hence the bond cannot form. The secondary structure of DNA is very robust because of the additive effect of many H-bonds between the base pairs, all oriented in the same way, i.e. stacked on top of each other. Figure 8(a) shows base pairing between adenine and thymine, and how it works for one portion of the double chain is shown diagrammatically in Figure 8(b). The helical tertiary structure is made possible by the orientation in space of the flat rings of the bases. If the double chain twists into a helix a large number of weak bonds can be formed between each ring and others lying directly above or below it. Again the sheer number of these bonds greatly enhances the stability of the molecule, but they are nevertheless individual, weak and easily broken. We shall see in later chapters how very important this is to nucleic acid function.

In three dimensions, Watson and Crick drew their model as in Figure 8(c). The whole concept brilliantly accounted for the X-ray analysis and also for the strange constancy of the ratios of the amounts of the different bases that had been so puzzling to Chargaff, who never forgave what he regarded as two rank amateurs for solving a problem which had foxed skilled biochemists and chemists for many years. In some ways the

Figure 8

(a) Mortice-and-tenon arrangement of base-pairs.

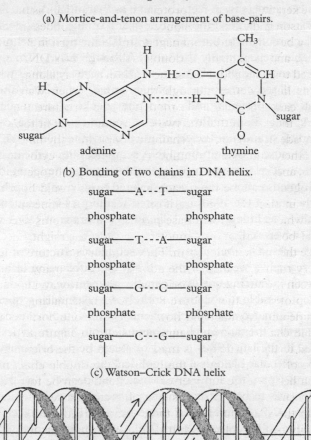

adenine   thymine

(b) Bonding of two chains in DNA helix.

```
sugar────A---T────sugar
  │                  │
phosphate        phosphate
  │                  │
sugar────T---A────sugar
  │                  │
phosphate        phosphate
  │                  │
sugar────G---C────sugar
  │                  │
phosphate        phosphate
  │                  │
sugar────C---G────sugar
  │                  │
```

(c) Watson–Crick DNA helix.

strange tensions that have existed between biochemists and molecular biologists for the getting-on half century since the Watson–Crick discovery can be dated to this episode.

In the structure of DNA shown here, the amount of adenine inevitably equals that of thymine, and guanine that of cytosine, just as the chemical analysis had shown. Despite the restrictions placed on the base order in such a model, the number of permissible isomers is, in fact, the same as that of RNA, that is, astronomically large. Finding the order of the bases presents similar problems to those faced with RNA, and the techniques available are essentially the same. Although DNA forms very long strands, individual lengths of DNA have sequences which, as we will see, correspond to particular lengths of RNA and ultimately to proteins, and when one speaks of sequencing a DNA, what is normally meant is sequencing such a corresponding strand, not the entire nuclear or mitochondrial material. Because understanding DNA sequences lies at the heart of the new developments in biotechnology and genetic engineering immense efforts have been devoted to solving the problem. By the end of the 1990s automated techniques for sequencing requiring as little as a few billionths of a gram of DNA were being advertised.

The result is that many DNA sequences are now known, and scarcely a week goes by without the publication of a new naturally occurring one (and genetic engineering methods now make it possible to construct entirely novel artificial sequences too). Automated sequencing machines and computer comparisons make the task more mechanical, though scarcely less tedious. In the late 1980s a head of steam began to develop in both molecular biological and political circles, especially in the US and Japan, for some grand biological project. The project chosen was to be the sequencing of the entire complement of human DNA – the Human Genome, or HUGO project. The task, due to be completed by 2003, was monumental, as the human genome consists of no fewer than three thousand million nucleotide base pairs. Even to write this sequence down in A-C-G-T format would require a book half a million pages long, and if the sequencing were to be done at the rate of one base pair a second without cease, night and day, would take up to 75 years to complete. Just what this project has so far achieved – and might in the future achieve – will be discussed later.

LIPIDS

We have left the lipids until last in this chapter mainly because they do not conform to our concept of a macromolecule quite as well as the other three classes already discussed. They are a diverse group, with few points in common other than the property of being insoluble in water but soluble in organic liquids – a very operational definition! They are much smaller than the other macromolecules, having molecular weights in the hundreds instead of thousands or millions, and their inherent structural hierarchy is a little more difficult to define. This is because they are rarely found as individual molecules, because they have a tendency towards self-aggregation, and it is the aggregate that we must consider as the macromolecule, the building blocks of the primary structure being the individual lipid molecules themselves. However, the building blocks in lipids are not held together by strong covalent bonds but by weak bonding due to interaction between their dominant non-polar regions. The absence of a strong backbone and the presence of only weak bonding in the structure of lipid aggregates means that the macromolecular property of flexibility is much more strongly emphasized, so that it can almost be described as fluidity. Lipids, as a group, include, as well as fats, esters, formed by the combination of an alcohol with an organic acid. The ester linkage is produced according to the equation

$$R.COOH + R'.OH \longleftrightarrow R.COOR' + H_2O$$

where R and R' may represent any organic radical.

Fats and oils

The prototype fat is a combination of a straight chain fatty acid, sixteen or eighteen carbon atoms long, with *glycerol,* a sweet, sticky liquid containing three alcoholic groups which can combine with one, two, or three fatty acid molecules, yielding a mono-, di-, or tri-glyceride.

An *oil* differs from a fat only in being liquid at normal

temperatures – that is, an oil is a fat with a low melting point.

The typical fat therefore has the formula:

$$CH_2OOCR$$
$$|$$
$$CHOOCR'$$
$$|$$
$$CH_2OOCR''$$

R, R', and R'' need not necessarily be the same, and this general formula can cover a variety of possible permutations. But in practice 80 per cent or more of fats contain one or more of the three acids:

palmitic $[CH_3(CH_2)_{14} COOH$
stearic $[CH_3(CH_2)_{16} COOH]$
oleic $[CH_3(CH_2)_7CH=CH(CH_2)_7 COOH]$

These three are simple straight chain acids; oleic differs from the other two in having a double bond. The fats are named according to the acids present, so that tristearin has three stearic acids, whilst if one oleic and two palmitics are found the fat is oleodipalmitin.

In a naturally occurring fat a mixture of many of the possible combinations will be found, although the exact proportions of each will depend on the source of the fat, and are not purely random; some combinations are preferred to others. In plants, double-bonded (*unsaturated*, or even *polyunsaturated*) acids tend to predominate – olive oil is almost pure triolein. Plants also have a higher proportion of the more unusual acids than do animals. In land animals, on the other hand, the single-bonded (*saturated*) acids are more frequent. This distinction is of considerable dietary interest as high concentrations of saturated as opposed to unsaturated fats in the diet can contribute – for reasons that will become clear as we proceed – to high levels of circulating cholesterol in the bloodstream, itself a contributory factor to coronary heart disease.

As the melting point of a fat is directly related to the

number of double bonds present in its constituent acids (the more double bonds, the lower the melting point) plant fats tend to be liquid at normal temperatures and are therefore called oils. If the double bond is saturated by hydrogenating the oil, a solidification ('hardening') occurs, an operation which is utilized in producing standard margarine by reducing vegetable oils with hydrogen. As concern about dietary levels of saturated fat has increased, so new techniques for making spreadable butter substitutes without saturating the plant oils have become popular.

Conversely, a fat left exposed to the air undergoes oxidation along the fatty acid chain, resulting in double bonds and hence liquefaction, which is familiar as the first stage of the complex process whereby butter goes rancid. Butter is a mixture of animal fats, and the 'rancidification' is continued by slight hydrolysis of some of the esters present, releasing free acids, amongst them the foul-smelling butyric acid. Another familiar smell, that of burning fat, is provided by the heat-decomposition product of glycerol, *acrolein*.

The main function of fats is, like that of glycogen and starch amongst the polysaccharides, to serve as a food store. Surplus food taken in by the animal is laid down as fat (adipose) tissue to be used again when harder times come. Such subcutaneous fat has other functions, too – it is believed to serve as an insulator against heat loss by the body, and also as a cushion for such delicate internal organs as the kidneys, which are generously embedded in fat. *Waxes*, which are lipid esters which contain rather longer-chain acids and alcohols than the fats, often have a role also as external protection, e.g. against water loss from the surfaces of insects, fruits, leaves, or petals.

Phospholipids

So far we have discussed neutral fats, but often a monoglyceride, instead of combining with other fatty acids, will react with the strongly electronegative phosphate group to give either monoacyl or diacyl phospholipids (depending on the number of fatty acids present). It is the reactivity of the phosphate group that enables phospholipid aggregates to form complex structures with other molecules, most notably with proteins in membranes. How the lipid aggregates form higher order structures depends on several variables: the degree of fatty acid substitution in a phospholipid, the presence or absence of strongly polar terminal groups, the degree of saturability in the fatty acids and the relative proportions of lipid an water in the mixture are all important considerations.

Diacyl phospholipid molecules are extremely well fitted for their role as membrane constituents mainly because they exhibit directionality in their relationship to H_2O. They tend to aggregate into bilayers, their hydrophobic tails, composed of the long fatty acid chains, are held together by weak bonds, whilst the strongly polar phosphate heads face outwards, enabling them to interact with the water that surrounds the membranes. We shall consider further aspects of membrane structure in the next chapter.

Neutral fats behave differently at similar concentrations mainly because they lack the strongly polar head so that their behaviour is determined by their non-polar residue, and in solution they tend to form small round fat droplets. It is this particular feature that makes them so suited as a food store. Monoacyl phospholipids like lysolecithin tend to behave more like the neutral fats and they prefer to form micelles even at quite low concentration. This means that the insertion of just a few lysolecithin molecules into a lipid bilayer (see the next chapter) will greatly disrupt its organized structure.

Steroids

Also included in the lipid group by reason of their solubility in organic solvents is a group of substances which seem at first sight quite different from the fats and phosphosolids. These are the steroids, whose many members are produced by a series of minor modifications on a basic pattern of seventeen carbon atoms linked together as four interlocking carbon rings:

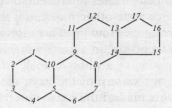

These rings, however, are simply the result of secondary folding of the long non-polar CH_2 chain, so the difference in structure is less great, though still important, than might at first appear.

A typical steroid is cholesterol, containing twenty-seven carbon atoms and an alcoholic group at carbon atom 3 of the steroid nucleus:

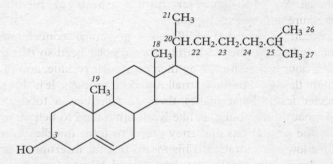

Cholesterol is a major constituent of the fatty insulating structures that protect nerves and is therefore present in very high concentrations in the brain. But it is also the starting point for the synthesis of many other important body constituents, including certain vitamins (e.g. vitamin D), several drugs and poisons, the bile acids important for digestion, and many hormones.

It is important to remember this because of the bad name that cholesterol has got in recent years as a result of the fact that derivatives of cholesterol itself also circulate in the bloodstream and form part of the walls of arteries and veins; this means that

THE CHEMISTRY OF LIFE

disorders in the body's use of cholesterol can result in the accumulation of cholesterol derivatives on the walls of blood vessels, thus 'furring them up', and are associated with arterial and coronary heart diseases.

This completes our survey of the large molecules of which living organisms are made; it is time to turn from them to a discussion of the structures which, in life, they compose and with which they interact – the internal organization of the unit of life, the cell.

CHAPTER 4

The Organization of the Cell

So far, we have treated living tissue as if it were merely to be regarded as a convenient source from which to extract various classes of chemical, using greater or lesser degrees of violence, but in any event, without regard for the structures from which those chemicals have been derived. This is essentially the chemists' view of living matter. For them, the cell is simply the logical extension of those chemicals we have so far discussed; a hierarchy runs from atom to molecule, from molecule to macromolecule with its primary, secondary, and tertiary structures. The cell is seen as a higher order of these molecules, its components an ensemble of quaternary structures, delicately related complexes of proteins, lipids, polysaccharides and nucleic acids, held in particular configurations by an appropriate ionic environment of water, sodium, potassium, and chloride ions. Implicit in this understanding is the belief, to which we have referred already, that the higher order structures and their interactions are essentially predictable from the lower order ones; that primary structure determines tertiary which in its turn determines cell structure.

A second way of looking at the cell is the physiologists'; for them, like other 'organismic' biologists, the cell is the smallest unit of which tissues and organs are composed – the lowest level of a new hierarchy; it is the interaction of cells which is of interest rather than what goes on inside them.

The third viewpoint is that of the biochemist or cell biologist, concerned both with what the cell is made of and how its structure is maintained; where the chemist's quaternary structures and the physiologist's smallest living unit meet. This is the terrain of the present chapter. In fact, in its earliest years, biochemistry got on by ignoring the cell, rather as we have done in the previous three chapters. The reasons were largely methodological. There was no way to relate chemistry to structure until the development of certain new tools of research in the 1940s – the ultracentrifuge and electron microscope. But the time has now come when we can legitimately stop pretending that the cell is made of smooth paste like butter in a butter-dish, and can recognize that it is more like marmalade – full of small, solid lumps, often quite different in texture and composition from the surrounding jelly.

The sciences long concerned with the microscopic study of the cell are *histology* and *cytology*; their relation to biochemistry is confirmed by two new subjects *histochemistry* and *cytochemistry* and to immunology by *immunocytochemistry*. The techniques of histology and cytology, developed over the last two hundred and fifty years, involve cutting very thin sections of biological material, mounting them on a microscope slide, staining them with a range of dyes that colour different parts of the specimen more or less intensely, examining them at various magnifications, and finally attempting to relate back the two-dimensional picture seen down the microscope tube to the original three-dimensional living tissue. By such techniques the early microscopists were able to study plant and animal tissues and to show that they were divided into a set of box-like compartments separated one from the other by thin walls. The organ, or the organism, they realized, was built of these boxes just as a wall is built of a set of individual bricks. They called the boxes, *cells* (apparently by analogy with monks' cells), and the thin walls, cell *membranes* (though plant cell membranes are themselves surrounded by a second line of fortification in the form of a more rigid cell *wall* of cellulose).

The cell is not merely the smallest unit of living organisms like plants and animals; for many forms of life it was found to be the *only* unit. *Moulds* such as yeast, and *bacteria*, all manage to compress the whole range of their activities within the membranes of one tiny cell – often considerably smaller even than the less-versatile plant or animal cells. Another familiar example of a one-celled animal is the little pond-water creature studied at one time or another by all biology students with access to a microscope – the *amoeba*. Early microscopists suspected, but

Table 4. Dimensions of molecules and cells

Species	Weight (g)	Atomic weight	Length (cm) (longest axis)
Hydrogen atom (H)	1.6×10^{-24}	1	1×10^{-8}
Glycine molecule (CH_2NH_2COOH)	1.2×10^{-22}	75	2×10^{-8}
Typical protein molecule	$3.2 \times 10^{-20} -$ 1.6×10^{-18}	$2 \times 10^4 -$ 1×10^6	$5 - 20 \times 10^{-7}$
Typical virus	$1 \times 10^{-16} -$ 2×10^{-15}	$4 \times 10^7 -$ 1×10^9	5×10^{-6}
Typical bacterial cell	8×10^{-7}		1×10^{-4}
Typical animal cell	$1 - 10 \times 10^{-5}$		$1 - 5 \times 10^{-3}$
Human	80 000		200

could not prove, the existence of even smaller organisms, the *viruses*, the detailed study of which depended on the advent of the electron microscope in the mid-twentieth century. The virus completed the range of 'living units' and is something of a special case, as we will see later.

Animal and plant cells are all rather similar in both size and construction. They are likely to be about one-thousandth of a centimetre in diameter, and to weigh up to about one ten-millionth of a gram. In a person weighing about 80 kilos, there might be as many as a million million (or 10^{12}) cells. Table 4 gives an idea of the range of sizes and weights with which one is dealing here. The cells from different species, say rats and humans, or algae and roses, differ from one another slightly in their properties, and so do the cells from the different organs of any one species. Thus liver cells are slightly different from kidney cells and both from brain or blood cells.

But the differences, as we shall see, are not nearly as great as the similarities, and for much of this book we shall be describing the biochemistry of an 'ideal' or 'typical' animal cell, which might be that of most organs of most species. Only very much later will we come to consider in any detail how the biochemistry of the cells of different organs and species actually differ from one another.

According to the first theories, cells were filled with a clear, jelly-like substance which was endowed with all the mystic properties of life, and was called *protoplasm*. By the 1830s, it had become clear that all animal and plant cells contained at their centre a large, oval body, darker than the surrounding protoplasm and occupying as much as one-third of the total cell area. This body was called the *nucleus* (and at a later date, of course, gave its name to the nucleic acids – see page 68). The nucleus is separated from the rest of the cell by a thin skin of its own – the nuclear membrane. By the 1880s and 1890s, the microscopists had shown that the protoplasm surrounding the nucleus was itself far from being a clear jelly. Instead, it is lumpy and granular, filled with small specks of material which the highest-powered microscopes revealed as taking characteristic shapes as minute ovals, rods, and spheres. These specks were called *mitochondria*.

But even the highest powered light microscope can enlarge only up to about 3000 times, and objects half-seen or only suspected at this magnification could not be identified until, in the last years of the 1940s, the ordinary microscope could be supplemented by a new sort which used not light at all but a beam of electrons to examine specimens, and which could produce magnifications ranging from 10 000 up to one million or more. At once, a whole world of substructures unsuspected by the light microscopist became visible. The nuclei and mitochondria could be examined in fine detail to show a complex internal organization of their own, and the remaining *cytoplasm* or *cytosol* (the terms which by now had replaced 'protoplasm' for whatever clear jelly was still considered to exist within the cell) was seen to be criss-crossed with a rich network of twisting strands, membranes, and small groups of tiny linked particles, the *ribosomes*, till it looks like a lace table-mat.

Meanwhile, techniques in light microscopy did not remain

unchanged; the old-fashioned dyes and stains of conventional microscopy have been supplemented by the recognition that many of the substances present in the cell have a natural fluorescence and when illuminated by light of one wavelength will themselves emit light of a different one; natural fluorescence can be supplemented by induced fluorescence and general histochemical stains replaced by specific fluorescently or radioactively tagged antibodies which enable precise location of particular molecules to be achieved; *confocal* microscopy means that instead of having to examine only very thin sections of tissue, the microscope can be focused down through thick slabs; cells can be isolated and grown in thin layers in dishes (*tissue culture*) to be studied individually; and the spinoffs from space and military computer technology have made available video-enhancement, 3-D and false colour reconstruction of images. The results are that today a microscopic presentation has something in common with a visit to a modern art gallery, full of abstract forms in breath-takingly beautiful colour. Some of the products of these new technologies are shown in Plates 1–8.

Figure 9 shows a much reduced, schematic drawing derived

Figure 9

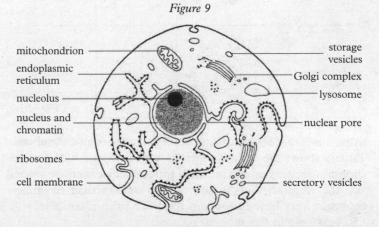

mitochondrion

endoplasmic
reticulum

nucleolus

nucleus and
chromatin

ribosomes

cell membrane

storage
vesicles

Golgi complex

lysosome

nuclear pore

secretory vesicles

Schematic diagram of a 'typical' animal cell

Table 5. Sedimentation of subcellular particles in a centrifuge

Particle	Shape	Sedimented after:	
		Time (min)	Gravitational force (g)
Nuclei		10	800
Mitochondria		15	12 000
Lysosomes		15	25 000
Ribosomes Membranes		60	105 000

from this naturally occurring beauty – a 'typical' animal cell. Table 5 shows the various particles that compose it; the mitochondria are fifty to one hundred times smaller than the nuclei and the ribosomes up to sixty times smaller than the mitochondria. Very little remains of the pure protoplasmic jelly of 150 years ago in this modern picture.

Appropriate questions for the biochemist to ask at this stage are: what are the individual subcellular structures composed of and how do they maintain their structural individuality? Are there chemical differences and specializations which distinguish them? It might be expected that certain activities are located in one organelle, others in another, and this localization is significant to an understanding of how the cell coordinates and regulates its activity. But to explore such possibilities there must be available some method of separating the organelles from one another, and obtaining on a relatively large scale (which means, for the biochemist, quantities in the order of thousandths of a gram – milligrams – of material) purified preparations of nuclei, mitochondria, ribosomes and so forth.

It is in the development of such methods that the techniques of *centrifugation* have proved invaluable. If the cell membrane is gently ruptured, all the internal organelles are released into suspension, like holding a plastic bag full of sand and water under water and puncturing it. Now, if the resulting suspension of sand is shaken up and allowed to settle, the densest and heaviest grains fall to the bottom first, then the medium ones, whilst some very fine grains remain suspended in the water even after several hours. The rate at which the particles fall depends on several factors: their weight, their density compared to the surrounding solution (a particle of cork, even a very large one, would not fall to the bottom however long one waited, as the density of cork is less than that of water, and it floats), and finally the gravitational force to which they are subjected. In our experiment with sand and water, the gravitational force is constant whilst the beaker holding the suspension is standing still, and the particles may take several hours to sediment. But ways exist of increasing the force towards the bottom of the beaker. Accelerating it upwards in a lift or rocket would be one way. Another would be to whirl it round very fast at the end of a long arm, when centrifugal force will tend to drive all the particles towards the part of the beaker furthest from the centre of the circle round which it is being rotated. Such a principle, called centrifugation, is used to simulate the increased force of gravity during rocket acceleration when training fledgling astronauts. It

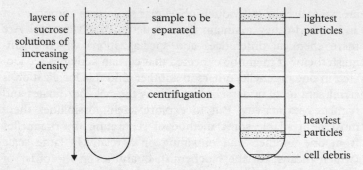

Figure 10. Sucrose density gradient centrifugation

is also the principle of the spindryer used in washing machines.

In the hands of the biochemists, centrifugation is a technique enabling them to separate out particles from a suspension of biological material rapidly and efficiently. Laboratory centrifuges are capable of spinning tubes containing nearly a litre of suspension at speeds of 60 to 70 000 r.p.m., which is the equivalent of applying a force of 500 000 times gravity to the particles in the suspension. By employing a carefully selected range of speeds and times, it is possible to sediment and remove, first the cell nuclei, the heaviest particles, then the mitochondria, the next heaviest, and finally the ribosomes and membranes. The clear solution which remains after all the particles have been removed represents the soluble or cytoplasmic fraction of the cell. One such selection of speeds and times is shown in Table 5. A refinement of this method is density gradient centrifugation. Here the suspension is layered on top of a series of sucrose solutions of varying concentrations so that the density of the liquid increases from the top to the bottom of the centrifuge tube. During centrifugation the cell organelles move through the gradient until they reach the sucrose concentration that has a density equal to their own, where they will pile up (see Figure 10). So at the end of the run, distinct bands can be seen at intervals down the tube, each band being a fairly homogeneous sample of one particular organelle. The accuracy and resolving power of this method are

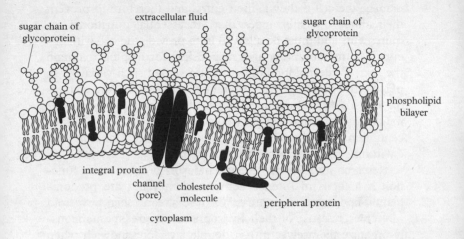

sugar chain of glycoprotein

extracellular fluid

sugar chain of glycoprotein

phospholipid bilayer

integral protein

channel (pore)

cholesterol molecule

peripheral protein

cytoplasm

Figure 11. The structure of the cell membrane

remarkable. Particles of very similar size can easily be separated in just a single spin instead of the many necessary with the differential method.

With this by way of background, we can now turn to a discussion of the individual subcellular components that can be seen under the electron microscope, isolated with the centrifuge, and studied chemically and biochemically.

MEMBRANES

Perhaps the best place to begin with our description of the cell is at its boundary, its point of contact with its environment, the cell membrane. The membrane has three main functions to perform. It has to maintain the structural integrity of the cell by its strength, insolubility and protective nature; it has to provide a surface of recognition which enables different cells to locate and attach to one another; and it has to act as a selective barrier to substances in the external environment, allowing the passage of some but not of others. The membrane has the power to discriminate between harmful substances that would damage the cell if they gained entry, and therefore to prevent their entry, and substances that are essential nutritional requirements and must be allowed in to provide the cell with energy. The membrane is also responsible for helping maintain the correct internal ion concentration so that such variables as pH, electric potential (the different concentrations and charges on ions inside and outside the cell) and osmotic pressure remain constant, often against wildly different external ionic concentrations that would tend to unbalance the internal environment if it were not for the membrane barrier.

Research into the precise mechanisms of membrane function is a focus of intense activity. Membranes are predominantly lipoprotein in nature. We have already seen how lipid molecules, because of their hydrophobic nature, spontaneously organize themselves into a double layer or sandwich when placed in a watery environment; their long non-polar tails line up together forming the cheese inside the sandwich, whilst their polar heads face outwards, where they can readily bind to protein or glycoprotein components which form the bread. The most widely accepted current theory of membrane structure is that of a core of lipid molecules arranged in a bilayer with its surface more or less covered with protein molecules, some of which penetrate right through the bilayer (see Figure 11). Because its secondary and tertiary structure is maintained by many weak bonds the lipid bilayer is very fluid and flexible; although again, because of the sheer number of internal hydrophobic bonds, it is at the same time strong and resistant to solvents. It is these characteristics of lipoproteins which give the membrane the capacity to carry out the first of its functions, that of maintaining the integrity of the cell.

The second function, that of recognition, seems to rely largely on the presence of the glycoprotein molecules embedded in the membrane surface. Glycoproteins, as we have seen in the previous chapter, are molecules in which polysaccharide sequences of glucose, galactose, mannose and other more complex sugars are attached to peptide chains. The peptide sequences can be envisaged as being embedded in the lipid membrane with the glyco-sequences pointing outward, frond-like into the external medium. These sequences can then match up with and stick on to complementary sequences on the surface of another cell, thus enabling cells to be assembled into tissues and organs; if the sequences don't match, the cells will not stick together but move past one another until they find an appropriate partner to recognize. A large class of various glycoproteins, called *Cell Adhesion Molecules* (CAMs), are involved in these recognition processes, which are of particular importance during the development of the organism, when it is growing fast and many new cells are being produced rapidly, needing to migrate to and find their appropriate site in the newly-forming organ. CAM-type glycoproteins are then often secreted by migrating cells, leaving snail-like trails for subsequent cells to move along.

Turning to the membrane's third function, there seem to be several ways for substances to pass through it to enter the cell. The simplest method, that works for the smaller molecules and ions, is one of *passive diffusion*, and occurs whenever their relative concentrations inside and outside the cell are different. Under these conditions a chemical gradient in the case of nonelectrolytes or an electrochemical gradient for charged particles will exist across the membrane. Molecules will move across the membrane from the region of high concentration to the region of low concentration until an equilibrium is reached, when diffusion will cease. Diffusion is easier the more soluble the substance is in the lipid of the membrane, and may be further facilitated by the presence of holes or pores in the membrane.

A method of *facilitated diffusion* that does not require much energy input makes use of the mechanism for transfer of Na^+ and K^+ ions that seems to exist within most membranes, the so-called ion pump. All living cells are polarized; the inside is electronegative with respect to the outside and this polarization is maintained by an unequal distribution of Na^+ and K^+ ions across the membrane. The importance of this polarization, too, will become clear in later chapters. The membrane pump is constantly shifting Na^+ ions out and K^+ ions into the cell so as to maintain a very pronounced ion gradient between the inside and outside of the cell. If a carrier protein in the membrane, together with its passenger molecule, complexes with a Na^+ ion, then the whole structure will move through the membrane by diffusion down the Na^+ gradient. Once inside the cell, the complex dissociates and Na^+ is expelled once more.

Active transport, on the other hand, is the process by which cells accumulate substances against a concentration gradient when the internal concentration of the molecules is already quite high and the external concentration is low. The molecule must then be helped across the membrane and it is specific protein molecules on the membrane surface which seem to be involved here. This class of proteins is called *transport proteins* – often glycoprotein in nature. They are able to recognize and clutch specific molecules out from the surrounding medium. They bind the molecule to be transported to themselves, and in so doing undergo a conformational change, carrying the molecule across the membrane, via a sort of protein bridge. The conformational change of the transporter protein that occurs during this process requires a driving force, which can be supplied by co-transport of ions, often Na^+. Binding mechanisms of this sort are not only vital in the transport of many molecules across the membrane but are also central, as we will see later, to the action of many hormones, to signalling processes between nerve cells, and to a multitude of drug effects.

The fourth function, that of communication between cells, relies on the presence of yet a further two classes of proteins – *ion channels* and *receptors*. Ion channels are membrane-spanning, like the transport proteins, but they mediate the flow of ions into and out of all cells and thus change the cell's membrane potential. In nerve cells, for example, the change in membrane potential alters excitability, as we shall see in Chapter 12. Receptors enable cells to respond to signals coming from outside the cell. Like transport proteins, they are glycoproteins which bind their target molecules, and then undergo a conformational change. However, rather than importing the bound molecule into the cell, the conformational change serves as a signal mechanism, triggering a cascade of intracellular events which mobilize further action – about which we will have more to say in Chapter 11.

All of these mechanisms are made possible by the marvellously fluid lipid environment which does not restrict either lateral or transverse movement within it, but nonetheless they do require an expenditure of energy by the cell. There will be a lot more to say about the source for this energy in subsequent chapters.

ENDOPLASMIC RETICULUM AND GOLGI APPARATUS

Membrane structures are not confined to the boundary of the cell. In electron micrographs the whole of the cytoplasm seems to be interlaced with a series of convoluted membranes forming a system of interconnecting channels (see Plate 5). This is called the *endoplasmic reticulum*; it consists of more than half the total membrane of the cell and is the route for the transport of material from one part of the cell to another. It also serves as an internal store for ions such as Ca^{2+}, releasing or trapping them as the need arises (see Chapter 11). It is thought to represent a continuous sheet, separating internal constituents in its *lumen* from the rest of the cytoplasm, although it is important to realize that the rather rigid, fixed internal structures which appear in electron microscope pictures may be misleading; in life, and in video sequences photographed from cells maintained in culture, the internal constituents of the cell are seen in constant motion, circulating through the cytoplasm, forming, breaking and reforming.

Also within the cytoplasm, visible in the electron microscope and separable by appropriate centrifugation procedures, is a system of disc-shaped, membrane-bound channels, three to twelve in number, stacked together to form the *Golgi apparatus*, so called after an Italian microscopist of the late nineteenth century (Plate 3). These structures are the sites of modification of proteins, in which the polysaccharide units are attached to them to form glycoproteins, and are particularly well developed

Figure 12

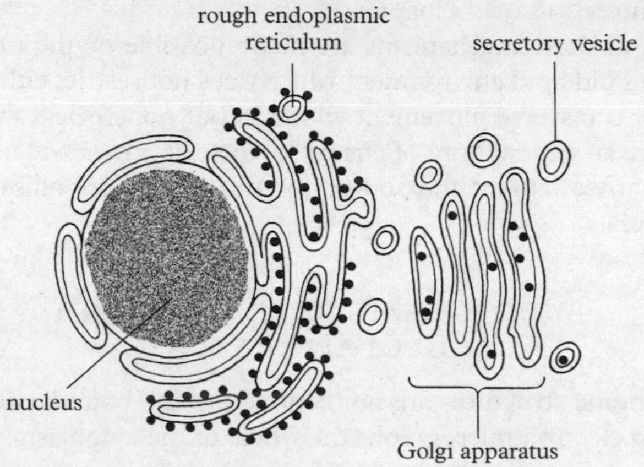

rough endoplasmic
reticulum   secretory vesicle

nucleus

Golgi apparatus

in cells which are specialized for the production and secretion of substances like hormones. They are involved in 'packaging' substances prior to transport, via the endoplasmic reticulum, from one part of the cell to another or for extrusion from the cell to the outside world.

NUCLEI

Of the organelles which can be separated by subcellular fractionation techniques, the largest and heaviest are the nuclei. A single nucleus is present in all animal and plant cells (with the exception of the red blood cells of some mammals) though not in bacteria, and it is intimately concerned with the reproductive mechanisms of the cell. Organisms with nuclei in their cells are known as *eukaryotes*. Those like bacteria without nuclei are *prokaryotes*. In eukaryotes, the nuclei contain virtually all the cell's DNA, a fair amount of RNA, and considerable amounts of basic proteins (histones, see page 69 and Plate 1) that we have already referred to as being associated with DNA. The complex of DNA/RNA/protein is called *chromatin* and at certain stages in the growth cycle of the cell, becomes organized into well-defined *chromosomes*, whose role will become clear later.

Very often found within the nucleus is a small densely stain-ing body called the *nucleolus*. The number of nucleoli present in any given nucleus varies in different cell types; they are packed with a particular class of RNA. Some RNA is also present in the nucleus in the form of ribosomes, although these are main-ly to be found in the cytoplasm (see below). The whole of the nucleus is constrained by a well-defined boundary, the nuclear membrane, which has the same type of lipoprotein bilayer structure as the cell membrane, except that it is perforated by some 3–4000 pores (gaps), each cmbedded in a large, disc-like protein complex; the pores are believed to allow the passage of RNA from the nucleus into the cytoplasm and proteins and other signalling substances from the cytoplasm into the nucleus (Plate 1). The significance of many of these structures and arrangements will become clearer when we discuss the mech-anisms of protein synthesis in Chapter 10.

MITOCHONDRIA

The next in order of magnitude of the cell substructures are the mitochondria (singular – mitochondrion). In many ways they are biochemically the most interesting, and have been intensive-ly studied over many years. Each cell may contain up to 100 or

Figure 13. A mitochondrion

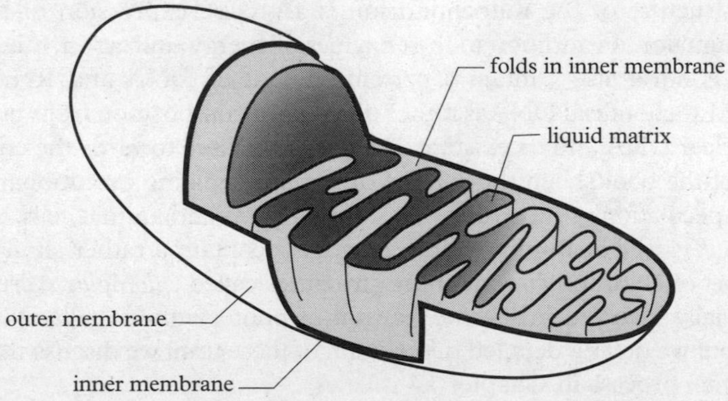

97

so mitochondria, each egg-shaped and between 1 and 4×10^{-6}m (10^{-6}m, or a millionth of a metre, is known as a micron: μ) in length. Each mitochondrion is surrounded by a double membrane, an inner and an outer one. Each membrane is a lipid bilayer. The outer membrane is smooth, but the inner is pulled into long folds that run across the entire width of the mitochondrion, so that in section the mitochondrion appears in the electron microscope like a pile of custard-cream biscuits. Each of the biscuits represents a fold of the inner membrane running across the mitochondrion, whilst between each fold is the custard cream, the concentrated solution with which the mitochondrion is filled. The folds of the inner membrane are known as *cristae*, the region between inner and outer membranes as the *intermembrane space*. The mitochondrial *matrix* is the compartment enclosed by the inner membrane (Figure 13, Plate 3).

In part, this membrane structure serves the same purpose for the mitochondrion as does the cell membrane for the cytoplasm, that is, of regulating the traffic of chemicals between inside and outside, but there is more to it than that. The mitochondrion, as we shall see in succeeding chapters, is the powerhouse of the cell, packed with enzymes involved in the production of utilizable energy by the breakdown of substances derived from food. Each of the different regions of the mitochondria, inner and outer membranes and the liquid intercristal space, has its part to play in this integrated process, and the structure of the mitochondrion is a precise expression of its function. In addition to the enzymes of energy utilization, mitochondria also contain a certain amount of DNA and RNA. Mitochondrial DNA is rather different in composition from nuclear DNA and its existence has, as we will see towards the end of the book, been the subject of some fascinating evolutionary speculations concerning the origin of plant and animal cells.

As well as mitochondria, plant cells contain a rather similar set of chlorophyll-containing structures called *chloroplasts*, crucially involved in the mechanism of photosynthesis (Plate 8), but we defer a detailed description of these until we discuss this vital process in Chapter 14.

LYSOSOMES AND PEROXISOMES

Next in size to the mitochondria comes a group of particles which, first discovered in liver by Christian de Duve of Brussels in 1953 but soon realized to be universal, provided a happy solution to an interesting 'teaser' which had for long intrigued biochemists. The cell, when broken up by homogenization, contains a variety of enzymes, whose function in assisting chemical reactions to occur we will discuss in the next chapter, which, between them, are capable of destroying nearly all the substances which go to make up the cell. How are they kept in check by the living cell and prevented from running amok? De Duve demonstrated that they were all contained within a class of particles he called *lysosomes* (Plate 2).

Under normal circumstances, the lysosomes are surrounded by a membrane which cages the enzymes in, keeping them from attacking other cell components. However, if the cell is damaged or dies, the lysosome membrane bursts, releasing the enzymes into solution and hence bringing about the rapid total dissolution of the cell. The lysosomes thus function as a party of cellular hyenas, harmless to the healthy, but quickly destroying sick or injured cells. On the other hand malfunctioning lysosomes, which may result from genetic disorder, are the cause of several types of disease and premature death.

Another group of somewhat similar particles are the *peroxisomes*, enzyme-containing structures, about 0.5μ in diameter and surrounded by single membranes; the enzymes within them are all concerned with the utilization and control of oxidation reactions by the cell.

RIBOSOMES

The smallest particles, the ribosomes, are only sedimented after prolonged centrifugation at high gravitational force. They are about 15×10^{-9} m (15 nanometres) in diameter, and the typical animal or plant cell contains about a million of them. Using the differential centrifugation method they are very difficult to separate from membrane components, and in electron micrographs they are frequently seen studded along the external surface of the endoplasmic reticulum, although others exist as free structures in the cytoplasm. There they are often found in multiples, linked together by a strand of RNA and somewhat like rosettes in appearance in the electron microscope; these ribosome assemblies are called *polysomes*, and they are most frequently to be found in cells that are particularly active in protein synthesis.

About half of the ribosome by weight is protein and half is RNA, and each ribosome is composed of two structural subunits, one large and one small, which fit together to form a complex with a molecular weight of several millions (see Figure 36, pages 246–7). The main function of the ribosome is in the manufacture of protein molecules from their amino acid building blocks and each of the subunits plays a different part in this process. We will discuss this in detail later. In many ways the ribosome can be regarded as a prime example of a quaternary, higher order molecular structure, for just as lipids and proteins in appropriate conditions will orient themselves so as to form a bilayered membrane, so too will the component RNA and proteins of the ribosome. As described in the previous chapter, if each ribosomal subunit is carefully fractionated it can be separated into thirty or forty separate protein and RNA fractions. If these fractions are then mixed together in approximately the right proportions, the conformational relationships between the individual RNA and protein molecules are such that they bind together and can be sedimented as re-formed ribosomes once more, capable of protein synthesis.

CYTOPLASM AND CYTOSKELETON

When all the particles have been sedimented out of suspension, there is left a clear solution which contains about thirty per cent of the total protein of the cell and all those substances which, in the intact cell, are not allotted to any of the membranes or particles but are presumed to remain free in solution in those spaces still left between the various particles and the reticulum.

The solution contains many of the simpler low-molecular-weight constituents of the cell, such as sodium, potassium, and phosphate ions, and also a number of enzyme systems with which much of the rest of this book will be concerned. However, it is probably a mistake to consider even the cytoplasm as a free solution; rather it probably exists for most of the time as a rather ordered molecular sol or gel, in which free mobility of components is strictly limited.

What is more, even the tourist guide to the cell we have offered above does not exhaust the types of visible structures present in the cytoplasm. For instance, the internal shape and structure of the cell are maintained by the existence of assemblages of thin fibres and tubes (*microfilaments, microtubules*) which radiate in all directions between the nucleus and cell membrane and seem to play an architectural role rather like the vaulting of a church roof or the struts of a tent, holding the entire cell in place; these filaments and tubules are of particular importance in cells which are not the standard shape of the 'typical' cell of Figure 8 but instead are long and thin, like nerve or muscle cells. These filaments and tubules form the cell's cytoskeleton. Up to twenty per cent of the protein of the cytoplasm can be involved in the cytoskeleton, one of the major constituents being a particular protein, tubulin, about 450 amino acids in length. Pure tubulin can readily be isolated from a mixture of cytoplasmic proteins and is of particular interest because, depending on the composition of the solution in which it is maintained and the presence of small quantities of other associated proteins, it can exist either in the form of individual soluble molecules, or can polymerize to produce quaternary structures in which arrays of tubulin molecules are stacked together to form tubules which look just like the microtubules present in the living cell – another example of the importance of self-assembly processes and of the interaction between the proteins and their microenvironment in the creation of structure (Figure 14). We need not go as far as the physicist Roger Penrose, who has argued that, at least in the brain, microtubules are the source of consciousness, to recognize their universal importance.

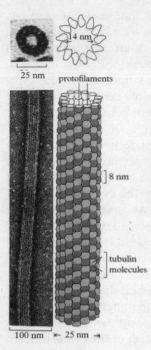

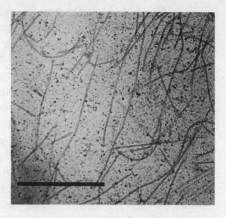

Figure 14. (a) Electron micrograph of a microtubule and its diagrammatic reconstruction. (b) Repolymerized tubulin forming microtubule-like structures. Scale bar 1 μm. 1nm=10⁻⁹ m

Even from this brief account, and before we really begin to come to grips with the nature of the various biochemical functions for which they are responsible, we can see that it is possible to regard the cell as a series of distinct compartments, each containing a multitude of particular small and large molecular weight components and involved in interlocking biochemical reactions. As we now come to consider these reactions in more detail, moving from the rather static picture of chemicals and cell structures – biochemistry as analysis – to the dynamics of biochemistry in action, it will become clear that this high level of structural organization of animal and plant cells – by comparison for instance with the much simpler organization of bacteria – has evolved as a most effective way of ensuring that such cells can regulate their own internal biochemistry in the interest of smooth and efficient running.

CHAPTER 5

Work and Enzymes

The biochemical world we have until now been describing is a static one; it is a world of things rather than processes, of photographs rather than cinefilms. In order to obtain this picture, we have had deliberately to destroy the living tissue at some arbitrary time, and systematically to fragment it into its components. Yet by doing so we have ignored just those characteristics which make life different from death. The proteins, nucleic acids, lipids, carbohydrates, of a living animal are to all intents and purposes identical with those of one just dead. But no one would mistake the one for the other. In this chemical world we have entered, in which our bodies have been reduced to chemical symbols and molecular architecture, where has the life gone? Can we describe the difference that we know exists between life and non-life in chemical terms?

Of course, we can. But to do so we must shift the perspective from which we have been studying the chemistry of the cell. We must no longer ask what it is made of but, rather, how is it kept in existence? For the highly organized organelles and macromolecules we have been describing all have one most important characteristic in common. By comparison with the simpler molecules they are all highly unstable. They readily begin to break down as soon as their environment alters beyond certain well-defined limits of pH and temperature. Thus, in the living system, any change that tends to alter these variables beyond their permissible limits must be opposed. We observed before how solutions of amino acids tended to resist attempts to change their pH by the addition of acid or alkali. The amino acid solution acted as a buffer against pH change. Similarly, the living organism is a mechanism which operates as a buffer against changes in its environment (we have already remarked on one of the buffer mechanisms, the selectivity of the cell membrane, and we will come across many more as we continue). When simple chemical systems operate to oppose change, they do so according to the principle of a French chemist, Le Chatelier; it was the French physiologist, Claude Bernard, who recognized that the same process also occurred in living organisms. Bernard spoke of the 'constancy of the interior environment' of the organism – a property now called *homeostasis*, and a central concept in biochemistry and physiology. Actually, however, it is something of a misnomer, as stasis implies a fixed, unchanging quality, and not merely are the cell and the organism in a state of dynamic rather than static equilibrium, but even this equilibrium is not stable but constantly adapting to the growth, development, activity and ageing of the organism. Perhaps *homeodynamic* would be a better term. Stasis is death.

The second characteristic implicit in the macromolecules is that they are all extremely unlikely substances. Those materials which life produces in such abundance still defeat the synthetic techniques of the chemist. In the living cell, such molecules cannot arise purely by random chemical reactions; they must be synthesized according to precisely planned pathways which can achieve a specificity far beyond that of the chemist. There must be mechanisms within the cell which can distinguish between even such close relatives as the (+) and (−) isomers of amino acids, or between sugars such as glucose and galactose.

This second problem would not perhaps be so overwhelming if the synthesis were, so to speak, a once-and-for-all job, if the lipid, protein, and carbohydrate molecules only had to be made in the desired quantity, laid down in the appropriate structure, and then were able to go on functioning indefinitely, until 'fair-wear-and-tear' demanded their replacement. But such is not the case. The use of radio-isotopes – radioactive versions of the normal molecules present in the cell or in foodstuffs – revealed that all body-components were in a constant state of flux; that protein, lipid, and nucleic acid molecules were constantly being renewed, old molecules being broken down and new ones synthesized to take their place. Even the molecules of such stable and unreactive tissue as bone and cartilage, which used to be regarded as quite inert, were found to have quite a short life-expectancy. Throughout the body, the molecule which survives for more than a few days without undergoing change was found to be the exception rather than the rule. The discovery of the prodigality of this constant flux of molecules, which leaves even the built-in obsolescence of such mass-produced items as cars and television sets standing, revolutionized biochemical thinking. It became clear that one, perhaps the, major function of the living cell was the constant re-creation of itself from within.

So before it can even begin to act on its external environment, the cell, and the living organism, has to provide the mechanisms whereby, first, it can protect itself against dissolution and destruction by the outside world, and, second, it can continuously resynthesize its more complex parts from much simpler molecules. These problems between them embrace that part of biochemistry concerned with kinetics.

How does the cell set about achieving the large number of highly complex chemical reactions necessary for the synthesis of macromolecules? And how does it obtain the energy required to accomplish its varied tasks?

For it is clear that the activities we have been describing represent work. To collect together several hundred amino acids at a particular site in the cell, to join them by means of peptide bonds, to coil and fold the resultant molecule, and to dispatch it at last to its appropriate site within the cell means that work has to be done upon the molecule. Similarly, to remove or render harmless poisonous or dangerous substances, and to keep the cellular pH within its desirable limits, requires work. And, of course, the moment the cell begins acting on its environment – the muscle cell by contracting, the nerve cell by transmitting impulses, endocrine cells by secreting their hormones – more work is done.

WORK AND ENERGY

The terms 'work' and 'energy' used in this sense may perhaps be unfamiliar ones. The concept of physical work is straightforward; we do work when we lift a weight from the ground, and the measure of the work done is represented by the mass we have lifted times the vertical distance through which it has travelled. Equally, the weight, once lifted, possesses potential energy.* When released, it will fall to the ground and, in falling, can perform work in its turn; it could be coupled to another weight over a pulley, for example, or used to drive a grandfather clock. Also, in physical terms, we are well aware of the law of the conservation of energy; we can get no more work out of a system than we have initially put into it. The grandfather clock will go on ticking until the weight has fallen again and all its potential energy has thus been used up. Then the clock will stop, and will only start again when the weight is lifted once more and hence more potential energy is pumped into the system.

This law of the conservation of energy makes it easy to understand the concept of work in a physical system, but it is not immediately obvious that work and energy are factors in chemical reactions too. Yet a battery-driven clock uses the chemical energy in the chemicals in the battery to achieve a similar function as the weights and pendulum. Consider the combination of carbon and oxygen to give carbon dioxide. We write the equation

$$C + O_2 \longrightarrow CO_2$$

but, in doing so, have omitted one important fact about it – that, during its course, about 420 000 joules† are released for every 12 grams‡ of carbon burned. In fact the equation ought

*Strictly, it is not the weight that possesses the energy, but the gravitational field acting upon it. When the weight is lifted, this energy is 'stored' in the gravitational field, which 'gives it back' when the weight falls.

†A joule is a unit of energy used when dealing with heat; it is approximately the amount of heat required to raise the temperature of 1 g of water through $0.24°$ C; there are 4.18 joules in the, perhaps more familiar, calorie. The term kJ (kilojoule) is used for a unit equivalent to 1000 joules.

‡This is the 'gram-atomic weight' or the atomic weight of carbon (which is 12) expressed in grams. Similarly, the gram-atomic weight of oxygen is 16 g and the gram-molecular weight of water is 18 g.

more accurately to take the form

$$C + O_2 \longrightarrow CO_2 + 420 \text{ kJ}$$

We can use these joules as a source of warmth, in a coal or gas fire, or can convert them into other forms of energy, as when oil is burned in an electricity generator. Thus the energy released in this chemical system is directly convertible to the energy we earlier described in a physical system – we could replace the falling weight of the grandfather clock, for instance, by driving the clock with electricity generated by burning coal. Equally, the law of conservation of energy holds in the chemical as well as in the physical system. When carbon reacts with oxygen the system releases energy, and in doing so itself loses energy. In order to convert the CO_2 back into $C + O_2$ again it would be necessary to supply exactly the same number of joules as those lost when CO_2 is formed.

The reaction $C + O_2 \longrightarrow CO_2$ is thermodynamically favourable and is described as *exergonic*, or energy releasing. The reverse reaction, $CO_2 \longrightarrow C + O_2$ is *endergonic* or energy-requiring. The energy change that occurs during such reactions, considered as occurring under standard conditions, is referred to by the measure $\Delta G°$, the standard free energy. In any reaction, if $\Delta G°$ is negative, the reaction is exergonic – that is, it can do work. If $\Delta G°$ is positive, the reaction requires energy input in order to take place.

Most of the reactions performed by the cell are endergonic; they require a source of energy. What is this source? Plants, as is well known, have a mechanism whereby they can tap the torrents of energy released every day by the complex atomic fusions which occur within the sun and pour down upon the earth as heat and light. The plants use the light energy for synthesizing sugars, such as glucose, and polysaccharides, such as starch, by the 'fixation' of CO_2. They can then break down the glucose and polysaccharides to release the energy once more when they require it. Animals have no primary energy source of this sort, and must rely on being able to find a ready-made supply of potted energy in the form of sugars. They achieve this by devouring the plants and using the materials which they have made. More sophisticated animals yet, of course, devour the animals which devour the plants, and the cycle of mutual interdependence is completed when the plants themselves utilize such waste-products as CO_2, which the animals release.

But the important thing about the whole cycle is that throughout it the law of conservation of energy holds. Human volunteers have, for example, been kept for several days in closed boxes, whilst the energy provided for them in the form of food, and that utilized – as heat output, carbon dioxide production, nitrogen excretion, and work done (for example, the riding of stationary bicycles) – are measured. Within the closest obtainable experimental limits, both humans and animals obey the law of conservation of energy, input and output being exactly balanced.

As with the body, so with the cell; and, in describing the many chemical syntheses and activities of the cell, we must at the same time ask also from where the energy for the performance of this work has come.

The principal source of energy for the cells is, of course, glucose. The complete burning of glucose goes according to the equation

$$C_6H_{12}O_6 + 6O_2 \longrightarrow 6CO_2 + 6H_2O + 2820 \text{ kJ}$$

The cell performs this reaction, but it does so in a rather roundabout manner. It breaks the glucose down by a series of reactions, each time releasing a small amount of energy only. The release of such a large amount of energy in one lump, so to speak, would be too much to cope with; much of the energy would be dissipated as heat, quite likely destroying the cell in the process. What the cell has been able to achieve for glucose is its controlled stepwise breakdown to provide a steady source of energy – just as an atomic power station is able to harness, control, and make (relatively) safe and useful the vast quantities of energy released as heat in a nuclear explosion.

Glucose is converted to CO_2 by a process involving nearly thirty different steps in each one of which a small amount of energy is released. Thus the cell obtains its energy in the form of small 'packets' which can be carefully conserved and used systematically. But it is important to note that in the final analysis the total amount of energy released during the glucose breakdown is exactly the same whether it is burned directly to CO_2 or passed through any number of different intermediate processes on the way. By way of illustration, consider the reaction

$$V \longrightarrow Z + 100\ 000\ j$$

Now suppose that the reaction instead occurs by way of a number of intermediates, W, X, Y. We can write the equations

$$Y \longrightarrow W + 25\ 000\ j$$
$$W \longrightarrow X + 25\ 000\ j$$
$$X \longrightarrow Y + 25\ 000\ j$$
$$Y \longrightarrow Z + 25\ 000\ j$$

Overall reaction $\quad V \longrightarrow Z + 100\ 000\ j$

By whichever route Y is converted to Z, the *total* energy change during the reactions is always the same. This is another consequence of the law of conservation of energy, and it means that the cell has nothing to lose, but a lot to gain, in its roundabout approach to free energy release.

We began by framing two questions with which to approach the world of biochemistry-as-kinetics: how does the cell achieve the chemical transformations that it undertakes, and where does it get the energy for so doing? We have now provided a general answer to the second question. But before we can proceed to answer it more fully, it is necessary to think again about the first.

ENZYMES

Few chemical reactions proceed completely spontaneously. Most need to be triggered off. Even thermodynamically favourable reactions, like the burning of carbon in oxygen that we considered in the last section, do not begin entirely by themselves. Carbon, as anyone knows who has tried to start a coal fire, is quite stable under normal conditions, and refuses to react with oxygen until its temperature is raised above a certain point. Once that point has been reached, the reaction goes ahead merrily until lack of carbon, or lack of oxygen, stops it. Until it has been reached, the carbon will sit indefinitely in the grate, refusing to burn. The amount of heat that has to be applied to the coal before it will catch fire represents the activation energy of the reaction. Once it is alight, the coal releases far more energy than had to be put into it to start it burning, but despite this, it would not begin without the added push. In this, the lump of coal waiting to catch light may be likened to a ball in a small hollow at the top of a long hill. (see Figure 15).

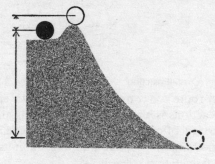

Figure 15

In order to release the potential energy present in the ball by virtue of its position at the top of the hill, a small amount of activation energy, in the form of a push, must be applied to it so as to roll it out of the hollow and over the top of the hill. Chemists can supply activation energy to their reactants without difficulty, by warming them in a test-tube, increasing the pressure on them if they are gases, altering their pH if they are liquids. The cell can do none of these things. The chemicals of which it is made begin to disintegrate if their temperature is raised much above 37°C or if their pH is shifted much from 7.0. How is it to provide the activation energy necessary for reactions to begin?

If all else failed, it could, of course, supply energy from its intracellular stores, but this would be a rather extravagant process. In some cases, as we shall find, that has to be done, for nothing else will suffice. But in most cases, a way round the difficulty has been found. What the cell does is to resort to low cunning rather than brute force. Another way of lifting the ball over the ridge of Figure 15 would be to go on rolling it along behind the ridge until a point was found where the ridge became sufficiently low for only the tiniest push to be needed to set the ball rolling down the hill. Chemically, this means that if we are trying to perform the reaction

$$A \longrightarrow B \tag{1}$$

which, although thermodynamically favourable, still requires a fairly high activation energy, we may instead look for another reaction, involving an intermediate X, which can react according to the equations

$$A + X \longrightarrow AX \tag{2}$$
$$AX \longrightarrow B + X \tag{3}$$

$$\overline{\text{Sum} \quad A \longrightarrow B} \tag{4}$$

of which the sum, reaction (4), is identical with the reaction (1) in which we were originally interested.

X is only of use to us if the activation energy of reactions (2) and (3) is so low as to allow them to occur spontaneously. In that case, reaction (4) will now also occur spontaneously as a result of (2) and (3). When this happens, we shall have achieved our purpose and found a point on the hilltop where the ridge was low enough to let the ball roll down without having to be heaved over the edge by the expenditure of much energy. It will be noticed that in the reaction scheme we have drawn, X, although taking part in the reactions, remains, at the end, unaltered.

Substances such as X are called *catalysts*, and the process of lowering the activation energy of a reaction is *catalysis*. Catalysis is a pretty common phenomenon in chemistry and finds much application, both in the laboratory and in commerce. Catalytic cracking of oil to give petrol and other derivatives, or the use of a special form of nickel as a catalyst in the reduction of oils, saturating their double bonds and turning them into fats such as margarine, are well-known processes.

Catalysis is practically universal in biochemistry, and the class of biochemical catalysts are collectively known as *enzymes*. We saw in the Introduction how they got their name when the brothers Buchner first extracted a preparation of yeast by grinding it with sand in a mortar and preparing a juice which could ferment sugar catalytically to give alcohol. A number of general rules govern the behaviour of enzymes, as of all catalysts. The sum of the rules is that enzymes can do only certain well-defined jobs, and can never violate the conventional laws of chemistry and physics, even though at first sight their incredible versatility might seem almost inexplicable.

(1) A catalyst can never catalyse a thermodynamically unfavourable reaction. It can never change an impossible into a possible reaction, but only make a possible but difficult one rather easier. It can never roll the ball uphill again once it has rolled to the bottom of the hill.

(2) A catalyst can never change the course of a reaction. If the reaction in the absence of the catalyst is A $\longrightarrow$ B, all that the catalyst will do is make A $\longrightarrow$ B easier. It won't make A $\longrightarrow$ C happen instead.

(3) Nor can a catalyst change the equilibrium of a reaction. If, in a reversible reaction

$$A + B \rightleftharpoons C + D$$

the reactants are mixed together and left for long enough, ultimately a characteristic and quite precise ratio of (A + B) to (C + D) will be achieved whether one started one's reaction mixture with A and B only, with C and D only, or with a mixture of all four. This ratio will be the same whether the reaction is catalysed by an enzyme or left to reach equilibrium by itself.

(4) In certain circumstances, though, a catalyst might exert a directing influence over a reaction. Thus, if the reactions

$$A \longrightarrow B$$
and
$$A \longrightarrow C$$

are *both* thermodynamically possible, a catalyst might help decide between them by lowering the activation energy required for one more than it lowers it for the other. This directive role can be very important amongst enzymes strategically placed at certain points along the pathways of metabolism followed within the cell.

(5) Finally, the general rule of catalysis, that although the catalyst may participate in the reaction it catalyses, in the end it is recovered unchanged from the reaction mixture. This is implicit in the equations (2) and (3) of page 111. It means that a very tiny amount of the catalyst can be used over and over again by a very large number of molecules of the reactant. Some enzymes, for example, are used and released again so rapidly that one enzyme molecule can catalyse the transformation of half a million molecules of the reactant per minute. Thus enzymes exert an influence on events within the cell quite out of proportion to the actual amount present.

Enzymes, then, are the tools the cell uses to manipulate, cut up and stick together the molecular raw material with which it is presented. One set of enzymes takes glucose down its long pathway to carbon dioxide, another set synthesizes proteins from amino acids or fats from fatty acids. Enzymes have, however, a special characteristic that distinguishes them from other, non-biological catalysts. Whilst chemical catalysts like nickel will readily catalyse a whole class of reactions of the same general type – that of reduction by hydrogen – enzymes are very highly selective about the reactions they choose to speed up and those they will not catalyse. They are not general-purpose but custom-built tools, and the cell requires a different enzyme for practically every reaction it carries out.

This specificity is not entirely absolute; for instance there are enzymes, such as esterases, which will catalyse the hydrolysis of an ester

$$R.COOR' + H_2O \longrightarrow R.COOH + R'OH$$

without much regard for the nature of the groups R, R', always provided there is an ester linkage between them. But other enzymes are slightly more choosy, and demand that at least one half of the molecule they act on be fixed; acid phosphatase, for example, catalyses the hydrolysis

$$R.OPO_3H_2 + H_2O \rightleftharpoons R.OH + H_3PO_4$$

Again R can be any one of many possible groups, but, if the phosphate half of the molecule is replaced by anything else, the enzyme at once refuses to work.

For the majority of enzymes, however, absolute specificity is the rule. For example, lactic dehydrogenase catalyses the oxidation of lactic to pyruvic acid

| lactic acid | pyruvic acid |

but does not act on lactic acid's near neighbour hydroxybutyric acid

$$
\begin{array}{c}
CH_3 \\
| \\
CH_2 \\
| \\
CH(OH) \\
| \\
CO_2H
\end{array}
$$

hydroxybutyric acid

What is more, not only is lactic dehydrogenase specific for lactic and no other acid, but it also discriminates between the two optical isomers of the acid, only accepting one form. Most enzymes are tailor-made like this to catalyse one and only one reaction.

As the cell performs many different reactions, this means that there must be a great number of different enzymes in any one cell. Thousands have already been identified, and more are being discovered all the time. Not all in the same cell of course, for many enzymes are involved in very specialized reactions relevant only to particular organisms – indeed a cynical enzymologist once remarked that given any reaction an organic chemist could suggest, it would be possible to find somewhere in the world of living things an enzyme to catalyse it. No one has yet proved this wrong. But many hundreds of enzymes are common to very large numbers of cells in species as widely different as carnations and mice, humans, mushrooms and sharks, so universal are the reactions they are called upon to perform. Indeed, as a result of their versatility in catalysing useful reactions that might otherwise prove very difficult to achieve, enzymes now have a regular place amongst the tools of industrial synthetic chemistry (quite apart from their addition to such domestic products as washing powders) and many substances, notably drugs and antibiotics, are produced with their help – a development which the expansion of biotechnology is bound only to increase. For, like any other catalyst, give an enzyme satisfactory working conditions and enough starting material to work with (the substance with which an enzyme reacts is called its *substrate*) and it will go on breaking down substrate and churning out reaction products practically indefinitely.

The first enzyme to be obtained purified and in crystalline form was urease (in 1926, by Sumner in America). Thousands have now been extensively purified and many isolated as crystals. All have turned out to be proteins (though some RNA molecules also do have catalytic activity and have been called ribozymes). Many enzymes also have non-protein material such as metal ions or nucleotides bound to them.

Enzyme kinetics

The kinetics of enzyme-catalysed reactions have been very intensively studied, not least because this is one of the few parts of biochemistry that can be brought under rigid mathematical

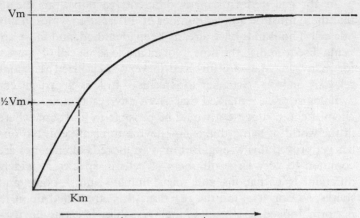

Figure 16. Michaelis–Menten plot

scrutiny and analysis. Many of the characteristics of enzyme action are restatements of the general laws of catalytic behaviour already summarized; most of the others can be deduced from the knowledge that enzymes are proteins.

Some of the key features of the behaviour of enzymes can be demonstrated simply by incubating an enzyme with its substrate and measuring the amount of product formed in a given time. From such a measurement, one can calculate the *rate* or velocity of the reaction, V (amount of product formed/time). If one performs this experiment several times with varying amounts of substrate, S, but the same amount of enzyme present, one can plot a graph of velocity versus substrate concentration (see Figure 16). Such a graph shows that as the substrate concentration increases, so does the velocity, at first, but it soon levels off to a point at which it cannot be increased by any further addition of substrate – this is the maximum velocity, Vm, which occurs when there is enough substrate present to complex with all of the enzyme. An estimate of Vm is therefore also a rough indication of the amount of enzyme present in a particular preparation. However, termination of the

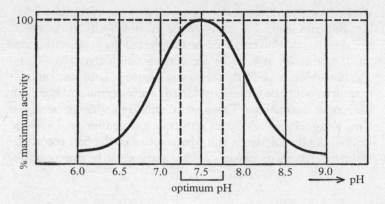

Figure 17. Enzyme activity and pH

reaction may also occur because the equilibrium-point of a reversible reaction has been reached and substrate and products are now present in balanced concentration, or because during the reaction there has been a change in pH, or the enzyme has been inactivated in some way.

For instance, the velocity of enzyme reactions is very much dependent on temperature. If the temperature is raised, the initial reaction velocity increases (as is the case for all chemical reactions, which go faster at higher temperatures), but also the enzyme is less stable and sooner inactivated. Most mammalian enzymes work best at 35–40°C, the temperature at which the body is normally maintained, whilst fish and other cold-blooded animals often have enzymes which function well at lower temperatures.

Their lability to heat is one indication of the protein nature of enzymes. Another is their sensitivity to pH. All enzymes have a particular pH at which they work fastest, and slight shifts in pH away from this point bring sharp falls in catalytic activity (see Figure 17). For most enzymes, the optimum pH is not far from neutrality, pH 7.0, although some, like the amylase of saliva (which is alkaline), work best at somewhat more alkaline pHs, whilst others, such as the pepsin of stomach digestive juice, function at the high acidity of pH 2 – 3. Hence in studying enzyme-catalysed reactions, one rarely looks at maximum velocity, but tries instead to measure the initial velocity of the reaction, i.e. the reaction rate over the first part of the graph of Figure 16. If one looks at a substrate concentration that gives half-maximal velocity (1/2Vm) (shown in Figure 16), this concentration turns out to be a constant for any one enzyme and is called the Km of the enzyme. The mathematical relationship between Vm, Km, velocity (V) and substrate concentration (S) was worked out by Michaelis and Menten, one of the first recorded women biochemists, as early as 1913 and is given by the equation

$$V = \frac{Vm. \ S}{Km + S}$$

Km is called the Michaelis constant and is a measure of the affinity of the enzyme for its substrate. The lower the Km, the higher the affinity. The important thing is that this equation holds true for most enzymes, so any deviations from the shape of the curve when V/S is plotted will tend to indicate, for instance, that the enzyme is not pure or that it does not have a high specificity for its substrate.

Although kinetic studies of enzymes and substrates studied in isolation from other cell constituents can only be an approximation of what is actually happening within the cell, the Michaelis–Menten equation, or derivations of it, are still extensively used, especially in clinical chemistry.

How do enzymes work? Like all catalysts, they are assumed to form a complex with their substrates which is then broken down to release the enzyme and the reaction products. Detailed study of the mechanisms of enzyme reactions and, in particular, measurement of the velocity with which the enzyme converts substrate into products under changing conditions, and with varying amounts of reactants present, confirmed this prediction theoretically long before it could be directly tested experimentally. But it is possible to provide direct experimental proof of this complex formation in some cases; for instance, where the substrate has a characteristic wavelength at which it absorbs or emits light (as do many organic substances), the formation and breakdown of the enzyme-substrate complex has been followed visually by watching the reaction and the changes in light absorption of the reactants in a spectrophotometer. Under certain circumstances, for a few enzymes, it is even possible to isolate the enzyme-substrate complexes and study them directly.

Knowing that a complex is formed between enzyme and substrate is one thing; demonstrating the precise details of the reaction of any one enzyme is another, especially as the enzymes themselves are proteins and their actual chemical structures often unknown. But despite the huge size of the enzyme molecule, the reaction that takes place involves only a small part of the amino acid chain, perhaps only half a dozen or a dozen amino acids in length, although in some enzymes the rest of the chain may play a part in controlling the reaction. This critical region is called the *active centre* of the enzyme and is vitally dependent on the higher-order structure of the protein. The twists and folds of a globular protein, with their multitude of weak bonds, bring portions of the amino acid chain together in space even though they may be separated by many intervening residues along the primary sequence of the chain. Thus pockets or clefts of a particular shape may be buried deep within the protein molecule (Figure 18) tailor-made to fit a particular substrate. The strength of union between enzyme and substrate is due to the formation of a large number of weak bonds between them, made possible by differences in electric charge distribution. Before these opposing charges can attract each other they have to be extremely close together. Thus the shape of the molecules at the point of attachment has to be exactly complementary to enable sufficient bonds to form. Any small mismatches will diminish the number of bonds that can be formed and the molecules will fail to recognize each other.

Thus the substrate fits into the enzyme like a key into a lock in such a way that the enzyme-substrate complex is much more reactive than the substrate alone. The enzyme may exert this effect in several ways. The substrate molecule may be distorted or strained as it becomes bound to the enzyme, so making it

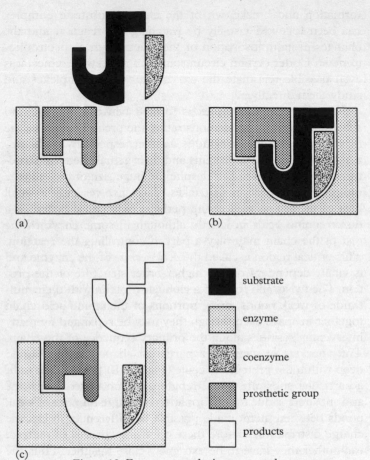

Figure 18. Enzymes, prosthetic groups and coenzymes

(a) Before reaction
Substrate and enzyme plus coenzyme come together.

(b) During reaction
All components locked together, thus activating substrate molecule.

(c) After reaction
Broken substrate molecule is released and enzyme is ready for further reaction.

easier to split; or, because of the precision of interaction, the enzyme may hold the substrate in exactly the right orientation for it to react with a second substrate molecule. The enzyme may also be responsible for concentrating substrate molecules in a localized area, therefore making it much more likely that they will meet and react with each other. This is especially true of the many enzymes that are firmly attached to membranes rather than freely wandering within the cytoplasm. As soon as the substrate has been transformed into the reaction products, they no longer fit into the space on the enzyme surface and are released into the surrounding solution, freeing the active centre for another substrate molecule.

In tailoring a shape to fit the substrate, the protein chain is frequently not adaptable enough, and the enzyme makes use of other substances to complete the pattern. Metal ions are amongst the most frequently used, particularly the ions of magnesium and manganese, both of which contain two positive charges and can thus provide a useful link between protein chain and negatively charged substrate. Sometimes entire compounds, such as nucleotides, are also bound to the enzyme surface to help the reaction, and often the enzyme is powerless to act without these additions.

The need for such extra factors in an enzyme reaction can often be shown by putting a solution of the enzyme in a sealed cellophane bag and suspending the bag and its contents in distilled water. The cellophane contains many minute pores, big enough to allow all low-molecular-weight materials to diffuse through them and out into the surrounding water, but small enough to prevent the large protein molecules from passing through. Thus after a few hours the bag will contain only the protein, and all the metal ions, nucleotides, and other low-molecular-weight substances will have leaked away. If the enzyme depended on any of them for its activity, it will now no longer function until they are added back again. This process of separation of high- and low-molecular-weight materials is called dialysis, and is frequently used to learn more about the way an enzyme works.

The substances which help an enzyme work are called *coenzymes*; frequently they are metal ions or such substances as nucleotides. Sometimes a coenzyme is bound so firmly to the enzyme surface that it is not even removed by dialysis; in this case it is described as a *prosthetic group*. But ions, coenzymes, and prosthetic groups all fulfil the same function of tailoring the enzyme to fit the substrate.

As more enzymes are sequenced, and information on the amino acid sequences around their active centres becomes available, it becomes possible to try to decipher general rules about them. It turns out that there are striking similarities amongst the active centres of various different enzymes within the same general class, such as phosphatases and esterases. Most notably, nearly all of these enzymes seem to possess at their active centres one or more serines, whilst other classes of enzyme make use of the sulphur-containing amino acid cysteine, suggesting that —SH groups play a role in complex formation in these cases. Because it is now becoming possible to synthesize new protein sequences more or less to order, it is also possible to create artificial enzymes, to test their catalytic properties and hence to make and test predictions about the important structural features of the active centre – quite apart from the prospect of making wholly new enzymes to carry out synthetic reactions that are uninteresting to actually existing living organisms but may be of considerable industrial significance. This possibility is creating a whole new science, of protein engineering, relying extensively on sophisticated computer technology to predict possible active centre configurations in theory before creating them in practice.

Enzyme inhibition

The extreme precision of the chemical tailoring at the active centre of the enzyme protein means that anything which alters the shape of the molecule at this point will interfere with the formation of the substrate complex. Enzymes which depend on a particular distribution of electric charge at the active centre will rapidly be inactivated by changes in pH which alter their distribution. Enzymes which depend on sulphydryl groups will be inactivated by any substance which reacts with —SH groups (as do salts or lead or mercury, for instance). Enzymes that depend on metal ion activation will cease to work when their activating ions are replaced by other metals, such as copper or iron. Substances which inactivate enzymes in this way are called *inhibitors*, and enzyme inhibition may or may not be reversible depending on the particular circumstances. If the process is irreversible, the enzyme is said to be poisoned. The term is no mere metaphor, for of course this is just the reason why many well-known poisons are indeed poisonous – they interfere with the workings of certain key enzymes.

Another form of inhibition is indicated when we remember that not only has the enzyme to be tailor-made to fit the substrate, but that the substrate itself must also precisely fit into the space available for it on the enzyme surface. If a large alteration in the substrate molecule occurs, it will no longer fit the space left for it and no reaction will occur. If it is only slightly changed, it may still be sufficiently like the original molecule to slip into the active centre, but, once there, will not be able to react as the original substrate would have done, and it will remain jammed in the active centre, stopping it from reacting with the genuine substrate. The process is exactly analogous to trying to open a door with a key that is just slightly wrong for the lock; we can put it in, but cannot take it out again, and the lock is jammed.

As both substrate and inhibitor can be regarded as fighting one another for the same active centre of the enzyme, this type of inhibition is called *competitive*. One example of considerable medical importance before the days of antibiotics is provided by certain bacteria, which need the substance para-aminobenzoic acid for growth. Traces of this substance, which occurs in blood and tissue, allow the bacteria to flourish. The closely related substance sulphanilamide, however, will compete with para-aminobenzoic acid for the bacterial enzymes, jamming them and hence preventing bacterial growth.

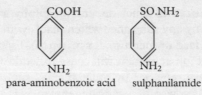

para-aminobenzoic acid   sulphanilamide

Hence the efficacy of the sulphanilamide drugs. Many other drugs and antibiotics have been discovered by using a similar principle, of finding an enzyme essential for the bacteria but not to the host tissue and then searching for a competitive inhibitor for it.

There is another class of inhibitor that has no affinity at all for the enzyme active site; in fact the inhibitor can only bind to the enzyme when it is complexed with its substrate. Thus this type of inhibitor works by preventing the splitting of substrates into products. Because there is no competition with the substrate for the active site, this is called non-competitive inhibition.

Allosteric enzymes

We have said that the active site of most enzymes only involves a small part of the protein molecule and that the major part of the amino acid chain has very little direct involvement in the catalytic action. But there is a certain group of enzymes where parts of the molecule far removed from the active site are essential for the functioning and control of enzyme activity. These enzymes were first recognized when a V/S curve was plotted for them, because the shape was not hyperbolic, like that of Figure 16, but sigmoid (like a stretched-out S). Thus this group of enzymes does not obey Michaelis–Menten kinetics. The sigmoidal shape of the V/S plot can be explained if the enzyme has more than one binding site for the substrate perhaps because the enzyme is actually composed of several subunits, each with its own active site. The binding of the first substrate molecule to one part of such an enzyme facilitates the binding of the second or subsequent substrate molecules by altering the shape of each individual region of the enzyme, a phenomenon known as cooperativity. The existence of enzymes with such properties makes it much easier for the cell to respond to a large demand for a particular product. Enzymes that have more than one binding site are called *allosteric*.

Allostery, however, is not confined to multisubunit enzymes. Some enzymes possess binding sites for substances other than the substrate and the effect of these substances may be to cause the enzyme molecule to change shape, thus facilitating or inhibiting substrate binding. For instance, in aspartate transcarbamylase, an enzyme that catalyses the first of a sequence of steps leading to the biosynthesis of pyrimidine nucleotides, ATP activates whilst its close relation cytosine triphosphate inhibits. Allosteric activation or inhibition of enzymes does not involve a substrate and an inhibitor binding for the same site, and is therefore non-competitive.

Two models have been proposed for the mechanism of allostery in enzymes. All known allosteric enzymes consist of two or more subunits, and the allosteric interactions are mediated by conformational changes transmitted between these subunits. In 1965 Jacques Monod, Jeffries Wyman and Jean-Pierre Changeaux in Paris proposed a model in which the enzyme could exist in only two states, called the tense (T) and the relaxed (R). The T and R states are in equilibrium. Activators and substrates favour the R state and shift the equilibrium towards it. Inhibitors favour the T state. In the absence of substrate nearly all the molecules are in the T form, but as substrate only binds to the R form, when it is added the conformational equilibrium is shifted towards the R form.

Although such a model is rather restrictive, it can easily account for the effects of allosteric activators and inhibitors. However, it is not the only possible mechanism. Daniel Koshland, the distinguished enzymologist so devoted to his subject that his autobiography is entitled *For the Love of Enzymes*, proposed an alternative, in which substrate, activator or inhibitor binding induces a conformational change in an enzyme subunit. A corresponding conformational change is then partially induced in an adjacent subunit contiguous with the subunit containing the bound substrate, activator or inhibitor. The effect of that binding is sequentially transmitted through all adjacent subunits, giving rise to an increasing or decreasing affinity for substrate by contiguous subunits.

In practice it may be that some cases of allostery use one of these models, some the other. In either event it has become apparent that very many of the subtle controlling influences on enzyme activity that are necessary for the smooth running of the cell, and that we discuss in Chapter 11, are mediated by such allosteric effects.

Enzyme classification

Whilst we will subsequently have much to do with enzymes in action and the reactions they catalyse, one further word ought now to be said about them. Despite the vast number of different enzymes and reactions that they catalyse, they can be classified as all conforming to a small number of types. In general, enzymes are labelled with the suffix -ase, preceded by the name of the substrate on which the enzyme acts. Thus esterases hydrolyse esters, succinic dehydrogenase removes hydrogen from succinic acid, and so forth. There are some exceptions, though; pepsin and trypsin, for example, and one or two other old-established enzymes have resisted most efforts to rationalize nomenclature. Enzymes are classified by grouping them together according to the type of reaction they catalyse. This gives each enzyme it systematic name, but within these major groups each enzyme is identified by means of a classification number plus a name for everyday use.

The six major subgroups are descriptive of the enzymes they contain:

Group 1: *Oxido-reductases*, of which alcohol dehydrogenase is a prime example, catalysing the reaction

$$C_2H_5OH \longrightarrow CH_3CHO + H_2$$
$$\text{alcohol} \qquad\qquad \text{acetaldehyde}$$

Group 2: *Transferases*, which catalyse reactions of the type

$$A - B + C \rightleftharpoons A + B - C$$

in which B is switched from A to C. A typical case is the synthesis of polysaccharides, when the enzyme maltase can start with two maltose molecules, each containing two glucose units joined head to tail, and convert them into triose and glucose:

$$\text{glucose—glucose} + \text{glucose—glucose} \rightleftharpoons$$
$$\text{(maltose)} \qquad\qquad \text{(maltose)}$$
$$\text{glucose—glucose—glucose} + \text{glucose}$$
$$\text{(triose)}$$

thus adding one extra unit to the growing polysaccharide chain.

Group 3: *Hydrolases*, which split an intramolecular bond by the addition of H_2O. They are typified by the digestive enzymes and several others such as the esterases which we have already mentioned.

Group 4: *Lyases*, which catalyse bond-breaking reactions other than those of hydrolysis, for example the removal of a group attached to a quadrivalent carbon atom to leave a double bond.

Group 5: *Isomerases*, which catalyse the molecular rearrangement of their substrates; the enzyme phosphoglucomutase, for example, lifts the phosphate off one part of a glucose phosphate molecule and fixes it to another part of the same molecule.

Group 6: *Ligases*, which catalyse bond-forming reactions, or the synthesis of a larger molecule by linking two precursor molecules.

Examples of all these types of enzyme in action will be found in the next few chapters.

CHAPTER 6

Pathways of Metabolism

The reactions catalysed by purified enzymes can be studied at leisure and with comparative ease. Most enzymes when purified are stable for quite long periods provided they are kept cold, and a biochemist who wishes to study the kinetics of an enzymic reaction can go to the laboratory in the morning, take a bottle of enzyme crystals or solution out of the freezer, mix it with the substrate, and analyse the reaction mixture at various times to see how much product has been formed. But, in the living cell, conditions are not so simple. There are many enzymes present, and the product of one enzymic reaction is itself likely to be the substrate of a second. Reactions occur very rapidly and within a very few seconds or minutes the molecules of a substrate may have been passed through the hands of a dozen or more enzymes, and have been transformed out of all recognition from the original starting substance. Molecules of glucose entering the cell, for example, are rapidly converted into an entire range of products. Some glucose is made into amino acids, some converted to fatty acids and then turned into lipids, some oxidized to carbon dioxide. Every one of these interconversions is carried out by a series of enzymes acting in sequence, and the whole procedure is as ordered and efficient as that of the mass production line in which sheet steel enters at one end and cars or washing machines roll off the other.

On the mass production line, in which each operation depends for its success on those that have gone before, so that a slight alteration in procedure or mistake at the beginning of the line affects everything that comes after, any interference in the operation of the line produces not a car or washing machine but a mass of tangled metal instead. Thus the biochemist, trying to follow and interpret the cellular production lines, all too frequently finds that in order to do so it is necessary to throw so large a spanner in the works that everything grinds to a halt.

But a knowledge of the route travelled by substrate molecules from beginning to end of their transformation by the cell is of great importance if we are ever to get a full picture of the way the cell works. So the effort must be made to draw up the 'flow sheets' for the chemical interconversions that the cell achieves. In this chapter, we consider some of the methods which can be used to draw up these flow sheets, or *metabolic pathways*, as they are known. We begin by discussing available techniques, and then show how these can be applied to particular 'preparations', as biologists tend to call the organisms – or parts of organisms – with which they work.

METHODS
Inhibitors

One effective and old-established method for identifying a metabolic pathway begins with the selection of a substance which specifically blocks one of a series of enzymes acting in sequence. Consider the reaction chain:

$$K \rightarrow L \rightarrow M \rightarrow N$$

where K, the added substrate, is converted by enzyme K-ase into L, which in turn is changed by L-ase into M and then by M-ase into N. What will now happen if we add an inhibitor which prevents the conversion of M into N by blocking M-ase? M will go on being made, but will not now be further utilized. So the amount of M present in the system will steadily build up. As its concentration increases, a time will come when the equilibrium of the reaction L $\rightarrow$ M begins to tilt in favour of L. When this happens the production of M from L will slow down and finally cease completely. But meanwhile enzyme K-ase will still go on converting K to L and hence the amount of L in the system will also begin to increase. Finally, K-ase will grind to a halt.

What is the effect of this? If we were to analyse the tissue in the absence of the inhibitor of M-ase, we should only have found K and N present, for L and M were used as fast as they were made. But with the inhibitor present, analysis will reveal the presence of L and M as well. So we know that L and M must be formed en route between K and N. It is as if we had wanted to study the make and type of cars and lorries going past on a fast road. Under normal conditions, they go past too quickly for us to be able to spot the make. But if we put in a traffic light on the road, the vehicles slow down and stop, and finally a queue of stationary cars builds up which we can examine at leisure.

Where can such inhibitors be obtained? In some cases it is possible to use specific molecules that, because of their similarity to the natural substrates, act as competitive inhibitors, and much chemical ingenuity has gone into formulating them. Another highly specific source of inhibitors has been derived from the procedure developed by Kohler and Milstein in Cambridge in 1975, for preparing specific antibodies to particular proteins, using techniques discussed in Chapters 13 and 15. Such antibodies (*monoclonal antibodies*) will bind specifically to particular proteins in a tissue preparation, and if the protein is an enzyme, it is likely to be inhibited. Another approach is to specifically prevent the synthesis of a key protein. As will become clear in later chapters, protein synthesis demands the production of specific RNA sequences, called messenger RNA. 'Antisense' strands of RNA prevent the message from functioning and hence stop the enzyme or other protein from being synthesized.

Study of the sequence K ⟶ N using such inhibitors shows that L and M are intermediates. We still do not know the order of the reaction. Either

$$K \longrightarrow L \longrightarrow M \longrightarrow N$$

or

$$K \longrightarrow M \longrightarrow L \longrightarrow N$$

might be possible. A combination of things must help us decide the right sequence. First, we may be able to isolate the enzymes K-ase and L-ase and show that K-ase produces L and L-ase makes M. But other enzymes might also exist; we don't know.

Then we may show that, in the presence of the inhibitor, M accumulates *before* L, indicating that M utilization is the first to be blocked, and L utilization only secondarily. Finally, we may use our knowledge of the chemical structures of L and M. We know that each enzyme catalyses only one reaction, and hence we may expect to find that chemically L is more like K, M more like N. It is like the party game in which we have to transform one word into another by changing one letter at a time, with the proviso that all the intermediate words also make sense. Suppose we have to get from BOAT to MEND and we know that our path must go via BENT, we can get there by way of the reaction pathway

$$BOAT \longrightarrow BEAT \longrightarrow BENT \longrightarrow BEND \longrightarrow MEND$$

and no other arrangement of the three intermediate words will fit the rules. Each enzyme can only change one letter in the word, and, if we read the rules of the game aright, it should be possible to put them in the right order. Examples of the use of inhibitors to study reaction pathways of this sort are common, and most of the metabolic pathways we discuss in later chapters have been studied by these methods.

Isotopes

The second method universally employed to follow metabolic pathways utilizes radioisotopes. Normal carbon has an atomic weight of 12, but there also exist other isotopes of carbon with atomic weights of 13 and 14 (written ^{13}C and ^{14}C). Like all isotopes, they are chemically identical to the 'normal' atom and take part in all the same chemical reactions. The one difference is that some isotopes of carbon such as ^{14}C are radioactive whilst the 'normal', i.e. the commonest form, ^{12}C, is not. Molecules of carbon compounds containing the radioactive ^{14}C in place of the 'normal' one can now readily be made. If we replace the ^{12}C atoms of glucose by those of ^{14}C, the glucose will be radioactive. If this glucose is now given to the cell to metabolize, all the substances into which the glucose is converted will begin to accumulate a proportion of the 'labelled'

^{14}C. Any substance that lies on the metabolic pathway from glucose will thus become radioactive. By analysing the tissue and identifying the radioactive compounds, we will be able to show into which substances glucose is, and into which it is not, converted. What is more, if we perform such analyses at various times we will discover the order in which they lie on the pathway. Take the reaction sequence

$$A—X \rightarrow B—X \rightarrow C—X \rightarrow D—X$$

and suppose we add a small amount (a 'pulse') of compound A labelled with radioactive $X°$ ($A—X°$). As soon as the reaction starts up, radioactive $B—X°$ will begin to be formed and the amount of radioactivity in $B—X$ will steadily increase. As more and more $B—X°$ is formed, $C—X°$ will begin to get radioactive also, and then $D—X°$. Meanwhile, as $A—X°$ becomes used up, or we flood the reaction mixture with a 'chase' of unlabelled $A—X$, the amount of radioactivity in $B—X°$ will slowly decline. Finally, all the radioactivity will be found in $D—X°$. A study of the time-course will immediately reveal the reaction sequence for which we are looking.

Radioactive isotopes used in this way are called 'tracers'. Apart from carbon, the radioactive isotopes of phosphorus and sulphur, and sometimes also the isotopes of nitrogen, oxygen, and hydrogen, are used in metabolic studies.

A more complicated problem arises where one particular compound is used as a precursor in the manufacture of many different substances. If the precursor is uniformly labelled, for instance if all its carbon atoms are in the ^{14}C form, many intermediates will subsequently be found to contain some of the radioactivity and it is very difficult to tell on which particular pathways these intermediates lie. However, it is also possible to specifically label one or two particular atoms in a precursor molecule – for example only carbon 1 and 6 of glucose – and as different parts of the degraded molecule proceed down different pathways, it becomes easy to see, using these particular tracers, exactly how the precursor is utilized. Although they have in many cases been superseded by more modern methods, the importance of tracer techniques since their introduction in the 1940s can scarcely be overestimated in revealing the 'dynamic state of body constituents' with which the whole of metabolic biochemistry is now concerned; Meselson and Stahl's famous 'pulse-chase' in the early 1950s revealed the semi-conservative replication of DNA; to Melvin Calvin in the 1960s they gave the clue to the initial stages of carbon dioxide fixation in photosynthesis; in the 1970s they were used to study the mechanism of the incorporation of 'labelled' amino acids into protein during protein synthesis; and in the 1980s the role of phosphorylation cycles and second messenger systems. Without them, a large part of our knowledge of metabolic activity would be impossible.

Classical isotopic techniques such as these, like the inhibitor, antibody and antisense techniques described above, are of their nature interventive – they demand the interference in some way with normally occurring metabolic sequences, and any interference of course alters the behaviour of the system that is being studied. (This is an aspect of what is known as the Heisenberg principle in atomic physics, but it is no less true for biological systems.) It would obviously be very desirable to have a method which could study metabolism without such interference, and another property – their magnetic characteristics – of the nuclei of many atoms makes this possible. In a strong magnetic field, the non-radioactive isotopes ^{13}C, ^{31}P, ^{15}N and others such as hydrogen (^{1}H) all give measurable signals which vary depending on the composition of the molecule of which the isotope is a part. This signal can be detected by *nuclear magnetic resonance* (NMR). The signals are not very strong, and only relatively abundant molecules can be studied in this way, but NMR techniques can now be applied to a range of tissue preparations from whole organs to tissue homogenates to detect rapid changes in concentration of key metabolic intermediates.

Electrodes and dyes

The methods that have been described so far are all applicable to rather large masses of tissue, containing at least many hundreds of thousands, even millions, of cells. Thus they provide information about the average metabolic changes going on within the cells. Yet this may be misleading, because the biochemical and metabolic properties of cells in any tissue are not all identical. Most body organs for instance are composed of different classes of cells each with specialized functions and very different biochemical properties, and one may well want to study these individual cells separately.

One way of doing so makes use of a modification of the isotope and inhibitor techniques discussed above. A radioactive isotope, for instance, can be injected into an animal and will be incorporated into various cells. Microscope slides of these cells can then be prepared, and coated with photographic emulsion. Radioactive emissions from the tissue will blacken the film in the region where the isotope is present in the cell, and can then be detected by developing the film and viewing it with a microscope. This process is known as *autoradiography*. Alternatively, an antibody can be given a radioactive or fluorescent tag, and localized in the same way, an imaging method known as immunocytochemistry. Specific RNA molecules can be located by an analogous technique, *in situ* hybridization.

But these are rather slow and static methods, and many of the changes one is interested in may occur within seconds or even fractions of seconds. Some of these have the effect of altering the electrical properties of the cell. We have already mentioned that there is a great difference in ionic concentration between the inside and the outside of cells, maintained by active transport and pumping processes across the cell membrane. This ionic difference can be measured by inserting a tiny, fine-tipped glass tube (a *micropipette* or *microelectrode*) filled with an electrically conducting salt solution into the cell. The cell membrane seals up again around the tube, and the voltage between the inside and outside of the cell can be measured. The tube can be used to inject minuscule quantities of ions or inhibitors into the cell, or by an appropriate choice of conducting properties can be used to selectively measure certain ions, for instance H^+ or Ca^{2+}, whose concentration and movement within the cell and across the cell membrane is, as we will see in subsequent chapters, a key metabolic event.

PREPARATIONS

Having described a number of available techniques, it is time to see how they can be applied. The simplest forms of metabolic experiments use whole animals. These are so-called *in vivo* studies, by contrast with those that use various forms of isolated preparation, or *in vitro* (literally, in glass, i.e. a test-tube or incubation dish). If an animal is fed on food of known composition, or containing radioactive tracers, and changes in the contents of the bloodstream, urine, and faeces are analysed, some indication of the fate and utilization of the food by the body can be obtained. But the potential of such a method is clearly limited; the information we can obtain is confined to the first and last stages of metabolism, and even the weight that we can put upon the results we do get is not certain. The conclusions we can draw are on the same level as those to be made about the activities and behaviour of a house full of people based on their grocery orders and the contents of their dustbins; revealing, but quite likely also misleading. But we can make such statements as 'when protein is fed to an animal, what nitrogen does not appear as faeces is largely recovered in the urine', indicating that the animal is in 'nitrogen balance', and neither gaining nor losing nitrogen.

The intact organism can be 'improved' for experimental purposes if it is rendered abnormal in some way, by genetic malfunction, by illness, or by operation. Some of the ethical issues surrounding such manipulation and experimentation on animals were discussed in the Introduction. Certain sorts of information can only, however, be obtained in this way. The classic example is the discovery of the role of insulin in diabetes, known for centuries as 'the pissing evil', in which large quantities of urine with 'the smell, colour and taste of honey' are produced. The urine contains glucose, which somehow instead of getting into the body cells, remains in the bloodstream and is eventually passed into the urine. But despite recognizing the condition, no effective treatment existed and patients simply died. In 1889 Oscar Minkowski and Josef von Mehring in Strasbourg found that if they removed the pancreas of a dog, the animal became diabetic. So could something in the pancreas alleviate diabetes, in dogs or humans? Simply mincing pancreas up and feeding it to diabetics didn't help. But in Toronto in the 1920s Frederick Banting and Charles Best tied up the ducts which lead from the pancreas in dogs and collected the juice that was secreted. Injected back into diabetic dogs or humans, the juice stopped the glucose being excreted, and instead enabled the cells to take it up; the patients recovered. Later Banting and Best were able to purify the active agent, insulin, from the pancreatic secretion. The story moved from being a laboratory curiosity into an effective medical intervention when Banting and Best's laboratory chief brought the US company Eli Lilly into collaboration to prepare large scale amounts of insulin from slaughterhouses – mainly from pigs. There are several lessons from this story but the one that concerns us primarily is simply that only the direct experimental manipulation of the dogs enabled Banting and Best to make the crucial discovery of the role of pancreatic juice and hence set about identifying and then purifying insulin. No *in vitro* method could have given this information.

Genetic defects, or *mutations*, are used widely in the study of bacterial metabolism, where they can be readily induced, for example through irradiation by X-rays or from a radioactive source. Genetic defects frequently reveal themselves – for reasons that will become clear in Chapter 10 – in the form of the absence of one specific enzyme, and metabolic studies with such enzymically defective preparations are of the same type as those made possible by the use of a specific enzymic inhibitor which we discussed above. Spontaneously occurring genetic defects in animals are rarer, but classic cases of the absence of specific enzymes and hence the accumulation of abnormal metabolites are provided in humans by the genetically carried diseases phenylketonuria and alkaptonuria. In both, unusual substances are excreted in the urine, and the analysis of the reasons for their appearance has led to valuable information about the mechanism of amino acid metabolism in the body. Thus in alkaptonuria the abnormal metabolite is homogentisic acid, and it was found to be no longer excreted if amino acids were excluded from the sufferer's diet; and, in particular, if the amino acid tyrosine was not present. It could then be shown that homogentisic acid lay on the pathway of tyrosine breakdown, and thus several steps in this pathway could be revealed.

New genetic techniques can go far beyond such 'accidents of nature', however. Genes responsible for the synthesis of specific proteins can now be deleted or added almost to order in laboratory mice and some other species. Such 'constructs', as the mutants are called – 'knock-outs' or 'knock-ins', depending on whether genes have been removed or added – can then be studied as they develop.

But despite such successes, the use of whole organisms to provide metabolic information is limited. 'Knocking out' a gene means that during development cells have to adjust many other aspects of their metabolism, increasing the synthesis of some proteins, decreasing others, in order to compensate for the deletion. If indeed they are to survive at all. Many such mutations turn out to be lethal. Again, what would be desirable would be a non-interventive method to measure internal metabolic processes in animals or humans whilst they were occurring. The nearest available approach to this at present comes from the development in medicine of scanners, and in particular the method entitled *positron emission tomography* (PET scanning). Like the other isotopic methods discussed above, these systems detect radioactive emissions, in this case positrons, released from isotopes which have a very short half-life, decaying completely within a matter of minutes to hours, and which can therefore be assumed to be harmless when injected. In these methods, the specially prepared isotopically labelled substance is injected into the bloodstream and an array of detectors placed around the head or body of human subjects or experimental animals. The emitted positrons can be measured and because of the ways in which the detectors are located can be tracked back to their source in particular tissue regions, so that computer maps of the distribution of the compound and its metabolites can be obtained, with a resolution of a few cubic millimetres. Such techniques are still limited in scope (the equipment and facilities for synthesizing the isotopes are extremely expensive), but the types of metabolic map they produce, especially in exploring such complex organs as the brain during various types of functional activity, can be profoundly interesting.

The next stage down from the use of the intact animal is to dissect out a whole organ – heart, or kidney, or brain – and to substitute for its normal blood supply an artificial system of arteries and veins carrying to and from the organ a substitute blood. Such preparations are called *perfused organs*, and can be kept alive, or functioning to all intents and purposes normally, for many hours, provided they are kept at body temperature and supplied in their artificial bloodstream with oxygen and glucose. By comparing the composition of the perfusing fluid entering via the arteries and leaving via the veins, much information can be obtained about the uptake and utilization of various substances by the tissue.

The development of such perfused systems has been accelerated by the obvious value of such techniques to surgeons wishing to isolate temporarily some of the organs of the body whilst performing difficult operations. They are of course also a step on the way to the development of the 'artificial' kidneys and hearts which are now in common use in hospitals. The biochemist can learn a lot from them. Thus the comparison of perfused liver and brain shows that there are considerable differences in the ability of these organs to take up and use glucose and amino acids such as glutamic acid. Recognition of these differences led to the finding that brain shows a tendency to convert added glucose into glutamic acid, as compared to liver, which either oxidizes its glucose or uses it to synthesize glycogen. Such differences provide revealing glimpses into the way different organs are biochemically specialized to carry out different physiological functions, but they still fail to supply some of the crucial information about, for instance, the exact pathways by which glucose is converted into glutamic acid or glycogen.

It is only at a simpler level of organization than the whole organ that the biochemist finds a system which can be freely manipulated and controlled, whilst still retaining enough of the characteristics of the organ of origin to be fairly sure that it is revealing a genuine and not a misleading metabolic route-map. If thin slices of tissue are prepared and rapidly transferred into a warm, oxygenated medium of suitable composition, the slices will continue for several hours to take up oxygen and give off carbon dioxide, at a rate not very different from that of the intact organ. This survival depends partly on the preparation of the slices, partly on the composition of the medium. With the medium, one is providing the tissue slice with an artificial environment that must contain many of the conditions of the real one if the slices are to survive at all. The most essential things about it are that it must contain a buffer to keep the pH at or about 7.4, small amounts of sodium, potassium, calcium, and magnesium ions, a source of substrate for the slice, such as glucose, and a steady supply of oxygen. No other organic substances are required at all; the slice will continue to behave in this simple medium as if it were still part of the organ from which it was cut and were still being provided with a regular supply of blood by way of its capillaries. Microelectrodes can be placed into its cells and their electrical properties recorded as if it were still *in vivo*. The only other requirements are that the slice is made rapidly on removal of the organ from the body, and that it is cut sufficiently thinly to make sure that the oxygen dissolved in the medium can get freely to all parts of it and enable its cells to breathe. The beauty of it is that the experimenter can precisely control all the variables of the experiment; change the composition of the surrounding medium, feed the slice any one of innumerable possible substrates alone or in combinations, start and stop the reaction at accurately determined times, and measure precisely how much oxygen has been used up, carbon dioxide produced, and products accumulated.

The detailed mechanisms of glucose breakdown, amino acid interconversions, and fatty acid synthesis were thus resolved in tissue slice preparations. Today they are used to tackle much more general problems of the regulation and control of the whole pattern of cell metabolism by hormonal regulators like insulin and thyroxine, and by such seemingly simple ions as potassium and calcium. Using the slices, it is possible to obtain a picture of the cell functioning as an integrated whole, with all its parts in harmonious interconnection. And it is this picture that is the theme of contemporary biochemistry.

At a level below the slice, techniques based on enzymic digestion of the tissue, or gentle mechanical disruption, can separate individual cells from one another, whilst leaving each cell relatively intact and undamaged. The mix of cells can then be directly studied, or the different classes of cells making up the original tissue separated from one another by centrifugation techniques. Mixtures of cells, or pure fractions, can be placed on to a glass or plastic surface in a small chamber bathed in a suitable medium somewhat richer in amino acids and essential cofactors than that required for the tissue slice. Under such conditions the isolated cells – a preparation known as *tissue culture* – will grow, divide and interact more or less as in real life. Tissue culture techniques for animal or plant cells owe much to methods originally devised for studying bacteria, and later for cancer research, as tumour cells have long been studied in this way. Cells can be maintained in cultures for many weeks, especially if they are derived from developing tissue from young or foetal organisms, and will often show remarkable properties of even higher-order self-assembly than those we have discussed within the cell; thus dispersions of liver, kidney or retinal cells maintained in appropriate cultures will grow, multiply and organize themselves in space to form what look like miniature, two-dimensional versions of the organs from which they were derived. Cells in culture of this sort are particularly suited to the use of microelectrodes, and have a special role to play in the production of monoclonal antibodies.

At the final level below the slice, the tissue may be completely disintegrated by homogenization in the laboratory equivalent of a kitchen mixing machine, or by grinding with a pestle and mortar. As has already been mentioned in Chapter 4, if the homogenization is conducted in an appropriately buffered salt or sucrose solution, the subcellular constituents may be preserved intact and separated by centrifugation. Different metabolic sequences can then be localized to particular cell compartments.

More robust homogenization will destroy the subcellular organelles as well, but will release into solution many enzymes whose activity cannot otherwise be easily studied because it is for various reasons masked in the slice or intact organ. Kinetics of individual enzyme reactions are much more easily followed in the homogenate than in the slice or organ, and for this reason the homogenate may be used to study the more detailed mechanism of a reaction first observed at a higher level of tissue organization.

A 'complete' statement of the biochemistry of the cell must not only be able to list what enzymes are present in the system, but where they are located within the cell, whether they are maximally active or lying dormant for lack of substrate or cofactors, and how their activity relates to that of all the other cellular components. The reasons for this will become more apparent when we turn to consider in due course the control of metabolism (Chapter 11). For the moment we should emphasize that each of the methods described above provides partial answers to partial questions. At the same time, each is likely to produce artefactual results. Inhibitors may have side effects on other enzymes or there may be otherwise unused sidepaths which become activated. The use of tissue preparations such as slices may alter factors like ionic environment or oxygen supply, or enable vital metabolites or cofactors to leak into the surrounding medium. Cell and subcellular preparations may introduce varying degrees of damage to the particles whose properties are being studied.

To examine the cell and its biochemistry it is necessary to interfere in some way with its normal functioning. Interpreting the results of this interference in terms appropriate to the 'reality' of the cell *in situ*, rather than the experimentally reduced artefact, requires the highest degree of critical awareness on the part of the experimenter. One must never forget that the substrate, enzyme, or cell is never really an isolate. Our experiments have to make it one in order to study it, but to interpret our findings, we must consider always the whole system. Metabolic biochemistry is fun, but it has its dangers!

From these general considerations we can now turn immediately to a description of some of the more important of the metabolic pathways that these methods have revealed.

CHAPTER 7

Energy-Providing Reactions

There is more than one way of classifying metabolic pathways. In older text-books, the division is normally made between pathways of the breakdown of already existing substances to simpler compounds – for example, the degradation of protein to amino acids and the subsequent breakdown and oxidation of the amino acids – and the reverse process, that of the synthesis of complex from simple substances – the synthesis of protein from amino acids. The breaking-down reactions are called *catabolic*, the synthetic reactions are *anabolic*, and the sum of all the catabolic and anabolic reactions occurring at any time represents the total of the cell's *metabolism*. Another way of looking at the same reactions though is to consider the standard free-energy change that occurs during their course, to divide them into exergonic or endergonic (energy yielding or energy using) reactions. However, overall, the cell – even assuming perfect efficiency – must be in energy balance. In practice of course efficiency is never perfect and energy is always being lost. This is why the cell never carries out the two types of reaction in isolation from each other; exergonic ones are always linked to endergonic reactions, and the one is used to drive the other. In any one pathway the two types of reactions are found to be linked by a common chemical intermediate, which possesses a highly reactive group. The intermediate with its reactive group is formed at the expense of one of a small number of compounds, ubiquitously present in the cell.

There are three main types of such small *group transfer molecule* which will appear in the discussions that follow: the amidine phosphates like creatine phosphate; the thioesters like coenzyme A (CoA); and the nucleotide phosphates. The most important of all of the group transfer molecules, and one of the most universally important biochemical molecules, is adenosine triphosphate, ATP (page 37). ATP is a highly reactive

143

substance which can readily donate its terminal phosphate group to other molecules, thus phosphorylating them. It can also act as an adenylating agent, in which case the whole of the adenine-ribose part of the molecule is transferred to another compound (page 36). In addition, certain macromolecules take part in energy and group transfer processes. Some proteins may be phosphorylated at their serine or threonine residues for instance; others (the so-called G-proteins) may be associated with ATP's close relative GTP (page 159), others still with CoA. The concept of energy coupling by group transfer molecules via common intermediates is so crucial to an understanding of how the metabolic pathways within the cell are ordered, that we must look first at the way ATP is generated and leave consideration of the way in which this ATP is used until a subsequent chapter.

The cell gets most of its energy by oxidizing foodstuffs. Oxygen is taken up from the air by way of the lungs, and enters the bloodstream to form a loose combination with the red, iron-containing blood pigment, haemoglobin. When the circulating blood reaches the body tissues, where the amount of oxygen is low, the haemoglobin-oxygen complex dissociates, releasing oxygen into solution, whence it can diffuse through the capillary walls and into the cells. In return, carbon dioxide produced by the cells passes into the bloodstream and is swept away to the lungs, where it is released into the air. Thus the cell is kept constantly supplied with enough oxygen to help produce the energy it needs.

Oxidations are reactions which tend to have a very high free-energy yield; we have already quoted the figure of 2820 kJ released during the oxidation of one gram-molecular weight of glucose. But so large an amount of energy must not be released all at once; instead, its release is the result of the sequential operation of a series of enzymes; together, they form the first of the metabolic pathways that we shall discuss in detail.

Let us first consider the nature of the oxidation reaction itself. In its simplest form such a reaction would be written

$$A + O_2 \longrightarrow AO_2 \tag{1}$$

where A, an oxidizable substrate, is oxygenated to AO_2. But, whilst the adding of oxygen to a molecule is one form of oxidation, there are others. One such is the removal of hydrogen, according to the reaction

$$AH_2 + X \longrightarrow A + XH_2 \tag{2}$$

In reaction (1) A was oxidized at the expense of oxygen. In reaction (2) hydrogen has been removed, that is, A has been oxidized, at the expense of another substance, X, which has been correspondingly reduced. Thus an oxidation has been performed in a reaction into which oxygen itself does not even enter.

There is also a third form of oxidation. In order to follow this, we must remember that the atom of hydrogen can be regarded as composed of the charged hydrogen ion, H^+ (a proton), plus one *negatively* charged electron, e^-. In aqueous solutions, where the ions H^+ and OH^- both exist, *the removal of an electron from a charged ion is, in many respects, equivalent to its oxidation*. Thus when the doubly charged ion of iron, the ferrous ion, is converted to the triply positive, ferric form, an oxidation reaction has also occurred:

$$Fe^{2+} \longrightarrow Fe^{3+} + e^- \tag{3}$$

All these reactions occur in the cell but, on the whole, the second and third seem easier to perform, and, what is more, can be performed without the direct intervention of oxygen itself. Reaction (1) does have some importance though. There are enzymes, called oxidases, which catalyse the direct oxidation of their substrates by atmospheric oxygen. They are found in most cells, but are especially common in plants. The effects of one of them, polyphenol oxidase, are familiar to those who peel their apples before eating them. Apples contain traces of a compound called catechol which is oxidized, under the influence of polyphenol oxidase, to a complex, dark brown substance. When an apple is cut its surface is exposed to the air and polyphenol oxidase can begin to work, turning the surface of the apple gradually brown.

145

But in general the cell prefers to use equation (2). In principle, the cell's oxidative apparatus consists of a series of substances like the X of that equation and also of the enzymes to catalyse the reactions. Substances like X are called *hydrogen carriers*, or, for reasons that will become clear shortly, *electron carriers*. The enzymes which catalyse the reactions are *dehydrogenases*. They are usually specific for the individual substrates that they oxidize, so there are dehydrogenases for every substance the cell can oxidize. We have already met lactic dehydrogenase, for instance (see page 114). On the other hand, the hydrogen or electron carriers themselves are not at all specific; the same set of carriers serves for practically all the oxidations performed in the cell, once the specific dehydrogenase has made the initial link-up by transferring hydrogen from A to X in the equation.

The electron carriers are arranged in a strict linear sequence which is maintained because they are all firmly embedded in the lipoprotein of the inner membrane of the mitochondrion within the cell. As this is also the site where energy-providing substances are degraded the whole process of hydrogen transfer from food to the terminal acceptor oxygen with the subsequent release of energy can take place quickly and efficiently. The carriers are mainly proteins carrying prosthetic groups in the form of metallic ions that are capable of being alternately oxidized and reduced. The measure of this oxidation/reduction capacity is termed their *redox potential*. Electrons are always transferred from substances with a high redox potential to those of lower potential. So this in part helps to explain the requirement for linear ordering of the carriers.

The electron transport process is started with a high potential energy substrate, AH_2, whose hydrogen or electron is passed down a chain of carriers, each time releasing a little energy, which can be utilized for the synthesis of a group transfer molecule such as ATP, until at the very end the last carrier transfers its H^+ to oxygen and thus disposes of it as a molecule of water. The enzyme for this last reaction is thus an oxidase:

$$ZH_2 + (O) \longrightarrow Z + H_2O \qquad\qquad (4)$$

The process works like one of the water-wheels formerly used for grinding corn. Water pours into the top bucket of the water-wheel at high potential energy, as it has some distance to fall. The first bucket breaks its fall and in doing so itself begins to move, carrying the wheel round. The water spills into the second bucket, yielding more energy and hence more movement, then into the third, the fourth, and finally into the stream at the bottom; during its fall down the series of buckets its potential energy has been tapped off at several points and used to spin the wheel. As with the water, so the hydrogen in its fall towards oxygen is checked at several points and its available energy tapped off.

The identity of most of the carriers is now well established. The main sequence is arranged in a series of three large multi-enzyme complexes embedded within the inner mitochondrial membrane, linked by smaller and less-firmly embedded carriers each with a redox potential midway between the value of the potentials of the complexes they are linking. The main components are:

(1) NAD-linked dehydrogenases
(2) Flavin-linked dehydrogenases
(3) Iron-sulphur proteins
(4) Cytochromes

The sequence is shown in Figure 19.

NAD

When the dehydrogenase enzymes were first studied, it was found that, if they were dialysed, they lost activity. A soluble co-factor was obviously leaking away, leaving the protein enzyme in the dialysis bag inactive. The cofactor was isolated and named Coenzyme 1, but was subsequently identified as substance X of our equations, the first of the hydrogen carriers. Obviously, in its absence, reaction (2) cannot proceed; the dehydrogenase has nothing to do with the hydrogens that it collects from A. Chemically, the cofactor turns out to be a nucleotide substance: nicotinamide adenine dinucleotide, or

Figure 19. The spatial arrangement of electron carriers and ATP synthase in the inner mitochondrial membrane. (a) A representation of the mitochondrial inner membrane. (b) Detail showing the relationship of the complexes of the electron transport chain.

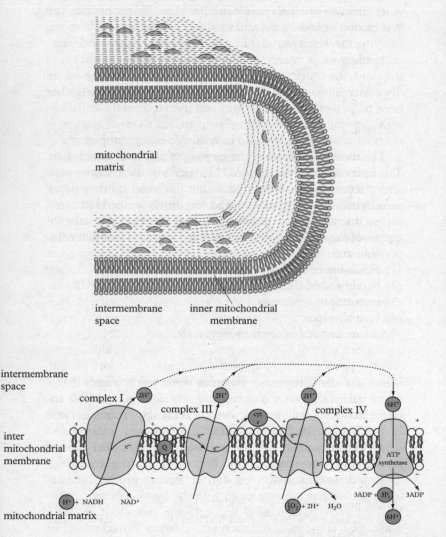

NAD* for short. We can write it to show its composition more clearly as:

nicotinamide-ribose-phosphate-phosphate-ribose-adenine

The business end of the molecule, that accepts the hydrogen, is the nicotinamide nucleotide, which reacts like this:

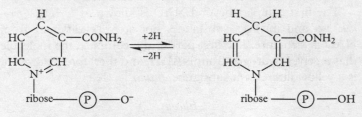

Nearly all the dehydrogenases pass hydrogen atoms on from their substrates direct to NAD. The hydrogen atoms arrive at NAD from many different points, like commuters descending upon a railway station, but, once having been taken on board, they all run the same common pathway to molecular oxygen at the terminus of the line. There is one branch line which enables the hydrogens from some substrates to bypass NAD completely, and there is also at least one other mainline to oxygen, which starts from a substance closely related to NAD – nicotinamide adenine dinucleotide phosphate, or NADP – the significance of which will become apparent in subsequent chapters. But these alternative routes are of minor significance compared with the NAD-line.

Linked with NAD are a group of iron- and sulphur-containing proteins, the ferredoxins, which form a sort of marshalling yard. If the next station along the line from NAD is temporarily full up, the hydrogen is passed between NAD and the ferredoxins until it can move on. All the hydrogens arriving at NAD then travel on through each of these main stations – a chain of three multienzyme complexes – in turn before arriving at the oxygen terminus. They do so very quickly; the time of transit down the hydrogen carrier line is in the order of

*More properly, this is written NAD^+, and its phosphorylated form $NADP^+$, and this is how you will find it in the textbooks. Here I have left it as NAD for simplicity.

thousandths of a second, and this rapidity, as well as the complexity of some of the chemical reactions involved and the fact that the reactions are coupled to the production of ATP, has made the study of the chemical nature of the hydrogen carriers very difficult, and the intimate details of the carrier mechanisms have only really been uncovered relatively recently.

The first complex beyond NAD is known as *NADH-Q reductase*, and involves substances not so very different from NAD in structure. The first part of the complex, the molecule that receives hydrogen from NADH, and therefore oxidizes it, is a yellow fluorescent substance, *flavin*.

Flavin

The flavin molecule consists of three linked rings:

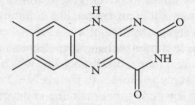

but the part of the molecule that actually does the work is the pair of nitrogens on the centre and right-hand rings, which can be alternately oxidized and reduced:

Like other biologically active substances we have met, flavin is found combined as a nucleotide, linked to ribose phosphate and sometimes also to adenine nucleotide, forming flavin mononucleotide (FMN) or flavin adenine dinucleotide (FAD). The flavin nucleotides, unlike NAD, are bound firmly on to the protein of the multienzyme complex which is therefore classed as *flavoprotein* (FP).

One other FP enzyme is also known, that which oxidizes succinic to fumaric acid:

$$\begin{array}{c} CH_2COOH \\ | \\ CH_2COOH \end{array} + FP \rightleftharpoons \begin{array}{c} CH.COOH \\ || \\ CH.COOH \end{array} + FPH_2$$

The existence of this enzyme complex, *succinic-CoQ reductase*, means that hydrogen atoms from succinic acid can be passed directly to the flavoprotein complex without travelling via NAD; this represents the one major branch line so far found by which hydrogen atoms can be dispatched to oxygen but bypass NAD.

Quinones

The FP complex in its turn can be oxidized by passing its hydrogens via a series of iron-sulphur prosthetic groups on to *ubiquinone*, or *coenzyme Q*. Quinones are molecules that contain two oxygen atoms bound to a 6-carbon ring, and can readily be reduced to *ubiquinol*.

In coenzyme Q, R is a long-chain fatty acid residue, but there

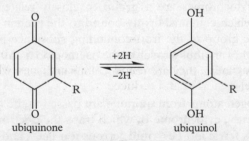

ubiquinone   ubiquinol

are many other related molecules in which R differs slightly – with only a little modification the molecule becomes that of vitamin E. It is also closely related to vitamin K and is now a familiar sight amongst the multivitamin shelves in healthfood stores, claimed to enhance energy metabolism. Coenzyme Q oxidizes reduced flavoprotein by accepting two hydrogens from it, and is subsequently oxidized in its turn by passing its hydrogens on to the next of the carriers, located in the next multienzyme complex, *CoQ-cytochrome c reductase*. With this complex we meet a major new class of molecules, the *cytochromes*.

The cytochromes

With the cytochromes, we move from a type of hydrogen transport in which hydrogen is moved directly by the alternate addition and subtraction of hydrogen atoms as in equation (2) of page 145, to a region of the carrier line in which the transfer is represented by alternate oxidation and reduction of an ion of iron, Fe, as in equation (3).

The hydrogen atom splits into a proton and an electron:

$$H \longrightarrow H^+ + e^-$$

and the electron is carried down through the remaining multienzyme complexes by means of the alternate oxidation and reduction of iron ions:

$$Fe^{2+} \longrightarrow Fe^{3+} + e^-$$

whilst the fate of the proton is intimately tied up, as we will shortly see, with the actual synthesis of ATP.

The cytochromes are a group of closely related proteins each of which contains, firmly bound to the protein surface, a prosthetic group of the iron-containing substance haem. The cytochromes are thus similar to the haemoglobin of blood, and, like haemoglobin, they are red in colour and can react directly with oxygen.

Hydrogen atoms from quinone are passed to the first of the cytochromes, cytochrome b, which traps the electron, using it to turn its ferric ion Fe^{3+} into ferrous ion Fe^{2+}, and parks the H^+ in the surrounding medium. The electron is passed from cytochrome b to the second and third cytochromes, c_1 and c. The ferric ion of c is reduced to ferrous, and the ferrous ion of b is oxidized to ferric once more. In the same way, the electron is passed from cytochrome c to cytochrome a and then to cytochrome a_3. The cytochromes are working like a line of basketball players, passing the electron ball from one to the other, and each time getting closer to the goal of oxygen. All of them, with the exception of cytochrome c, are firmly bound to the multienzyme complexes of the inner mitochondrial membrane; cytochrome c can readily be extracted – though when it is, the carrier line of course ceases to function.

With cytochrome a_3, the oxygen goal is reached at last. This cytochrome differs from the others in that as well as possessing the iron haem group it is also closely associated with the copper ion Cu^{2+}. Copper, like iron, can exist in the cuprous Cu^{2+} and cupric Cu^{3+} forms; the transfer of electrons to molecular oxygen occurs from the copper ion after it has been reduced by the Fe^{2+} of cytochrome a_3. So the final reaction in the electron transport chain involves two electrons, two protons and an oxygen atom reacting together to form water:

$$2Cu^{2+} + 2H^+ + O \longrightarrow 2Cu^{2+} + H_2O$$

Because the cytochrome a_3 complex works by oxidizing the earlier cytochromes at the expense of oxygen, it can be regarded as an enzyme, *cytochrome oxidase*.

Incidentally, the oxidation of reduced cytochrome a_3 is prevented by cyanide, and it is this that accounts for the poisonous effects of cyanide in the body. If cytochromes cannot be oxidized, hydrogen transport cannot occur and ATP cannot be made. The result is a quick death through energy-lack. Some narcotic drugs, on the other hand, work by slowing down the rate at which oxidized cytochromes can be reduced by accepting electrons from further back in the hydrogen transport line, resulting in a small but significant drop in available energy levels.

We can now draw in all the stations of the hydrogen/electron transport line, in a sequence running from substrate to oxygen, including the branch line from succinate which bypasses NAD, leaving us only to resolve the main question of how this process is coupled to ATP synthesis:

OXIDATIVE PHOSPHORYLATION

The point of sending the hydrogen atoms down a series of carriers on their way to oxygen is to tap off the available energy, so we may expect to find that their progress down the line is coupled with the production of ATP. Such indeed is the case, and, for oxidation to take place via the carrier line, it is necessary for ADP and phosphate to be present. During oxidation these substances are steadily used up, and ATP formed. When all the ADP or phosphate is finished, oxidation comes to a halt. If oxygen is not present, the cytochromes become reduced and the system will back up to the point where there is excess of reduced coenzymes which cannot find enough available substate on to which to discharge their hydrogens. The coupling mechanism is thus rather like the linkage between the engine of a car and its wheels, by way of the gears. Oxidation causes the engine to turn, and this in turn spins the wheels and forms ATP. But, if we stop the wheels from turning by putting on the brake or removing the ADP supply, the engine stalls and oxidation ceases. In a car, we can prevent this stalling by uncoupling the engine from the wheels an letting it tick over freely by putting the gears in neutral. The cell too has such a mechanism for uncoupling oxidation from phosphorylation, which we will discuss later when we come to consider how the cell regulates its activities. But normally the two are firmly linked.

Three questions now need to be answered:

(1) Why ADP?
(2) Which are the coupling points along the electron transport chain for ATP production?
(3) What is the mechanism of coupling?

The first two questions are relatively straightforward. Adenosine, it will be remembered, is the nucleotide formed from adenine and ribose, and it can be phosphorylated to yield first adenosine monophosphate (AMP), then adenosine diphosphate (ADP) and finally ATP (see page 36). If we follow the course of its hydrolysis back to adenosine we find the following series of reactions:

$$ATP \xrightarrow{H_2O} ADP + \textcircled{P} \qquad \Delta G° = -31 \text{ kJ}$$

$$ADP \xrightarrow{H_2O} AMP + \textcircled{P} \qquad \Delta G° = -31 \text{ kJ}$$

$$AMP \xrightarrow{H_2O} A + \textcircled{P} \qquad \Delta G° = -12.5 \text{ kJ}$$

Thus, while the first two reactions are highly exergonic, the hydrolysis of the third and last phosphate is much less so. In this, the hydrolysis of AMP resembles the majority of phosphate esters. It is the di- and triphosphate bonds of ATP which are exceptional. The important thing though is that all the reactions are fully reversible; the synthesis of ATP is thus highly endergonic and provides a convenient method of using the sometimes vast amounts of energy released from glucose oxidation. During the sequence of oxidation steps leading from glucose to oxygen, whenever a point is reached where a reaction occurs with a $\Delta G°$ of more than -31 kJ, this reaction can immediately be coupled with the synthesis of a molecule of ATP from ADP and phosphate. The ATP thus formed can then be utilized, at perhaps a different time and place within the cell, in the way we already mentioned at the beginning of the chapter.

If a quantitative study is made of the amount of ADP and phosphate used and ATP made per pair of hydrogen atoms passed down the line, it is found that three molecules of ADP disappear and three of ATP are formed during the transport of hydrogen from $NADH_2$ to oxygen. A clue as to the coupling points is provided by the observation that the number of ATP molecules formed during the oxidation of succinic acid is not three but two. As the succinic acid and NAD lines merge at the quinone stage, and are identical from then on, it follows that the extra ATP made on the NAD line must be formed between NAD and quinone, whilst two molecules of ATP are made between quinone and oxygen. More refined analytical techniques involving the isolation of submitochondrial complexes carrying different parts of the carrier chain, the sophisticated use of substances which inhibit particular carriers, and artificial electron donors that can only feed into the chain at specified points have pinpointed the other two sites as lying between cytochrome b and cytochrome c, and between cytochrome a and O_2.

The answer to our third question, that of the mechanism of coupling, was for many years a mysterious and controversial issue, until resolved by a radically novel hypothesis advanced by Peter Mitchell in 1961. Before that time, it was assumed that there ought to be chemical intermediates between the carrier chain and ATP, but none were ever observed. By contrast, there were a number of established experimental results, which did need to be accounted for. The most important of these are:

(1) Coupled ATP synthesis occurs only on intact mitochondrial membranes which surround a completely enclosed space.
(2) The enzyme responsible for ATP synthesis, ATP synthase, is a huge multiunit protein molecule and is firmly bound to the inner mitochondrial membrane. It can also in certain circumstances act in reverse, to hydrolyse ATP to ADP and ⓟi; that is, it can function as an ATP-ase.
(3) The energy released during electron transport is not always immediately used for ATP synthesis. It can somehow be stored to be used later, even after transport has stopped altogether.
(4) During the electron transport process, movement of protons across the mitochondrial membrane has been shown to occur.

Mitchell's chemiosmotic hypothesis, first greeted with much scepticism, came by the 1970s to be generally accepted, so much so that he was awarded the 1978 Nobel Prize for this work. It postulates that the electron carriers are arranged so precisely, spanning the inner mitochondrial membrane, that protons can be pumped from the inner mitochondrial matrix to the space between inner and outer membranes. In addition to the pumping of protons as a consequence of the oxidation-reduction of the carriers, during the reactions there is a conformational change in the membrane, which results in the ejection of protons from the matrix to the other side of the membrane, thus producing an electrically charged proton gradient across it. Just as energy can be stored in batteries because of the separation of the positive and negative charges within it, energy can be generated as a consequence of the separation of charges in membraneous system of mitochondria.

The enzyme catalysing the synthesis of ATP (ATP synthase) appears in electron micrographs as a series of spherical projections on the matrix side of the inner mitochondrial membrane, and is also present in chloroplasts in photosynthesizing green plants and some bacteria. In recent years a considerable effort has been devoted to purifying the various components of the mitochondrial ATP synthase – an achievement which was rewarded with a Nobel Prize in 1997. This enzyme is a large multicomponent complex made of at least 8 different subunits with a suggested molecular weight of 500 000. It contains a hydrophilic portion (F1) and hydrophobic, membrane-bound portion (Fo). ATP synthase uses energy stored in a proton gradient, which forms a *proton-motive force*, to produce ATP from ADP and $\textcircled{P}$i The hydrophobic portion of ATP acts as a proton channel and is responsible for proton transport, while F1 is the site of ATP synthesis. The conversion of ADP + $\textcircled{P}$i into ATP occurs spontaneously on the F1 portion of the enzyme but the release of ATP from the enzyme into the medium is the major energy-requiring step. The enzyme behaves almost as if it were a minute spinning wheel in which ion flow through the membrane portion of Fo generates torque that is transmitted between the enzyme's subunits, causing the F1 portion to rotate, expelling ATP into the mitochondrial matrix (Figure 20). It takes the translocation of $3H^+$ through the synthase to synthesize 1 molecule of ATP.

The synthesis of ATP by way of a transmembrane proton gradient is now recognized as but one of a number of examples of the importance of such gradients in biochemical processes and in the mechanism of transfer of ions and small molecules from one cellular region to another. The mechanism, which at

Figure 20. The ATP synthase enzyme

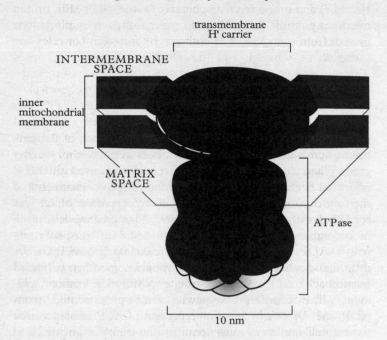

first sight seems complex, is in reality simple, requiring only a thin, closed insulating lipid membrane between two watery, proton-conducting solutions. A gradient of protons generated at one point along this membrane can be tapped off at another point quite a distance away, making it a mechanism relevant, as we will see, to a multitude of biological systems.

Other group transfer syntheses

The syntheses of all the other group transfer molecules we mentioned at the beginning of this chapter are, like that of ATP, highly endergonic. Creatine phosphate (CrP) has the structure

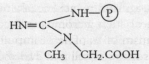

Like ATP, creatine phosphate can be hydrolysed to yield creatine (Cr) and inorganic phosphate, with a $\Delta G°$ of $-31kJ$. It then becomes possible for ATP and creatine to transfer phosphate groups from one to another without either needing or releasing energy.

$$A- \text{(P)}- \text{(P)}- \text{(P)} + Cr \rightleftharpoons A- \text{(P)}- \text{(P)} + Cr- \text{(P)}$$

This reaction too has a part in the energy balance of the cell, for the creatine phosphate can serve as yet a further energy store. When all the available ADP has been converted into ATP, and what Fritz Lipmann, the discoverer of this mechanism and the first during the 1940s to realize the central role of ATP in the cellular economy, called the 'energy-bank' of the cell is full, ATP begins to transfer phosphate to creatine phosphate, releasing ADP once more to be converted again into ATP. On the other hand, when ATP levels are low, phosphate can be transferred back from creatine phosphate to ADP, replenishing ATP stocks. If, in Lipmann's somewhat financially-oriented metaphor, the ATP-ADP system represents the current account of the cell's energy-bank, creatine phosphate is the deposit account.

Just as creatine phosphate can be synthesized from ATP, so can other of the nucleotides, such as guanosine triphosphate (GTP), and the phosphorylated proteins, without squandering the energetic advantages locked up within ATP's terminal phosphate bond. We can thus describe the methods whereby the cell gets the energy it needs. Many different reactions can provide ATP. As many different ones subsequently draw on the energy locked up inside that molecule. Much of the rest of this book will be occupied by describing some of the ways by which ATP is made and used. We shall find ATP being made during glucose, fatty acid, and amino acid breakdown, and ATP being split in the synthesis of macromolecules, the contraction of muscle and the transmission of nerve impulses, the transport of wanted substances into the cell and undesirable ones out of it.

Every phase of the cell's activity is dominated by the availability of this universal energy currency.

In its absence, the cell runs down, ceases to be able to perform its physiological functions, and in desperation begins to break open its last remaining food reserves. When ATP is plentiful, the cell relaxes, slows down the oxidation of glucose, and uses its superabundance of energy to build up long-term storage reserves of fats and carbohydrates. If any substance deserves to rank as central to biochemistry-as-kinetics, then it is ATP.

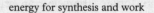

energy for synthesis and work

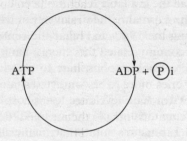

ATP   ADP + (P)i

energy from food

Figure 21

CHAPTER 8

Sources of Energy

FOOD

The substrates which the cell oxidizes in the manner we described in the last chapter are food. A unicellular organism, like amoeba, swallows its food whole – the cell literally engulfs particles of nutrient and sucks them into itself. In more complex animals, however, the food reaches the cell only after it has already been broken down into relatively small molecules which the cell can easily absorb and handle. This process of breaking down is called *digestion*, and its details are well known. In the human, food is broken down in the mouth, stomach, and intestine by a series of enzymes – amylases in the saliva and pancreatic juices for carbohydrates, pepsin in the stomach and trypsin and chymotrypsin in the intestine for proteins, and lipases in the intestine for fats. The resulting mixture of sugars, amino acids, and fatty acids is absorbed through the intestinal wall, and, directly in the case of the sugars and amino acids, indirectly in the case of fats, enters the bloodstream. In the well-fed individual, the cell is thus kept steadily supplied with its food requirements as low-molecular-weight substances dissolved in the circulating blood from which it can draw at will. Under normal circumstances the bulk of the cell's energy requirements are met by the breakdown of sugars, although, as we shall see, both fatty acids and amino acids can also, if the need arises, be broken down and the energy they release used to synthesize ATP.

GLUCOSE METABOLISM

The glucose circulating in the body's bloodstream serves to keep the cells supplied with a steady source of potential energy. Many cells, notably liver and muscle, absorb from the blood more than they require for their immediate needs, and this surplus is turned into glycogen, to guard against harder times later on. If the glycogen stores are already full, the liver cells begin to convert the surplus glucose into fat, which can subsequently be deposited in storage depots in various convenient portions of the body. It is this primitive defence mechanism against the possibility of glucose-starvation that has results so unflattering to the vanity of those humans fortunate enough to be able to eat each day more glucose-providing substances than their bodies really need; but for other animals, and for the large proportion of the world's population that exists precariously on the borders of starvation, it is an absolutely essential mechanism for not wasting any potential supplies of energy. We shall consider the detailed mechanisms of the synthesis of fat and glycogen in the next chapter. For the moment our concern is with the fate of the glucose immediately utilized as an energy source.

The breakdown of glucose is accomplished in two stages. In the first, the 6-carbon glucose molecule is split into two 3-carbon fragments of pyruvic acid, releasing only a little energy in the process. This first half of glucose breakdown can be carried out in the absence of oxygen, and is called *glycolysis*. In the second stage, the pyruvic acid is completely oxidized to carbon dioxide and water by way of a series of acidic intermediates. This stage releases a great deal of energy, is oxygen-requiring, and is called *glucose oxidation*.

When a situation arises in which the cell needs a great deal of energy very quickly, it may happen that the demand for oxygen is so great that it outruns the supply available to the cell from the haemoglobin of the bloodstream. When this happens, the cell falls back on operating the glycolytic half of glucose breakdown without the oxidative half. The energy yield from this is smaller, but it does mean that the cell has a second line of defence should its oxygen supplies fail it. Glucose breakdown in the absence of oxygen is called *anaerobic* glycolysis. It results in the production of pyruvic acid which the cell reduces to lactic acid and then discharges into the bloodstream.

Anaerobic glycolysis provides the energy for any sustained bout of violent muscular exercise – a hundred yards sprint, for example – and, indeed, after such exercise it is possible to measure large increases in the level of lactic acid in the athlete's bloodstream. The glycolytic pathway is also that followed in many bacteria and also in yeast when it ferments sugar to alcohol.

Glycolysis then is a practically universal pathway of glucose degradation, and has been studied for a long time in very many organisms. Much of the pioneer work was done with yeast, because of the exceptional interest that alcohol production has always had for scientists and non-scientists alike; it was only later that it was found that the same sequence of reactions which had been elucidated in yeast operated also in human muscle. The full discovery of the reaction sequence took many years, beginning with the Buchners in Germany and Harden and Young in England at the beginning of the twentieth century. The details were fully worked out by Embden and Meyerhof in Germany in the 1930s, and the glycolytic pathway is thus often known after them as the Embden–Meyerhof pathway. The oxidative part of the pathway, leading from pyruvic acid to carbon dioxide, was more difficult; it was not fully understood until the mid 1950s.

As we might expect from what we already know about enzyme reactions, the cell goes a long and laborious way about the breakdown of glucose. No less than ten enzymes are required to take it down as far as pyruvic acid, and another ten are needed in the subsequent oxidation of pyruvic acid to carbon dioxide and water. The enormous importance of this reaction sequence makes it necessary for us to try to describe it in full here, particularly as it will also serve us as a model for many later reaction sequences which we shall mention but not consider in anything like such detail. For the reader in a hurry, a judicious turning over of the pages until Figure 22 (page 171) and from there to Figure 23 (page 174) would be permissible. But for the stronger-willed, here is the fuller story.

GLYCOLYSIS

The first step towards the breakdown of glucose is at first sight a strange one; it demands the *phosphorylation* of glucose. The fact is that in order for glucose to participate in the reactions that follow, it must be activated at the expense of ATP. Fortunately the ATP is later recovered, with dividends. The phosphorylation is carried out by the enzyme *hexokinase*:

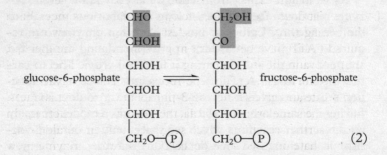

$$
\begin{array}{ll}
\text{CHO} & \text{CHO} \\
| & | \\
\text{CHOH} & \text{CHOH} \\
| & | \\
\text{CHOH} + \text{ATP} \longrightarrow \text{CHOH} + \text{ADP} \\
| & | \\
\text{CHOH} & \text{CHOH} \\
| & | \\
\text{COOH} & \text{CHOH} \\
| & | \\
\text{CH}_2\text{OH} & \text{CH}_2\text{O}-\boxed{\text{P}}
\end{array}
$$

glucose   glucose-6-phosphate   (1)

Although this reaction consumes ATP, it serves to trap glucose as its phosphate ester, whose charged nature prevents it from diffusing back across the cell membrane.

Glucose-6-phosphate is then converted by the enzyme *phosphohexoisomerase* to fructose-6-phosphate.

$$
\begin{array}{ll}
\text{CHO} & \text{CH}_2\text{OH} \\
| & | \\
\text{CHOH} & \text{CO} \\
| & | \\
\text{CHOH} & \text{CHOH} \\
| \rightleftharpoons & | \\
\text{CHOH} & \text{CHOH} \\
| & | \\
\text{CHOH} & \text{CHOH} \\
| & | \\
\text{CH}_2\text{O}-\boxed{\text{P}} & \text{CH}_2\text{O}-\boxed{\text{P}}
\end{array}
$$

glucose-6-phosphate   fructose-6-phosphate   (2)

A further phosphate group is then transferred from ATP to produce the highly reactive molecule fructose-1, 6-diphosphate, catalysed by the enzyme *phosphofructokinase*.

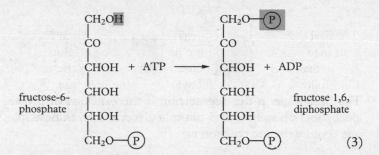

In the fourth reaction, the fructose-1, 6-diphosphate is split into two 3-carbon fragments by the enzyme *aldolase*:

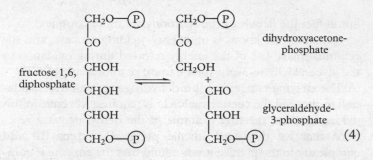

Thus in four separate reactions, the glucose molecule has been split down the middle into two separate substances, each containing three carbon atoms. At the same time, two molecules of ATP have been spent in phosphorylating the glucose, thus activating it and preparing it for breakdown. The stage is now set for these ATPs to be recovered. Dihydroxyacetone-phosphate and glyceraldehyde-3-phosphate are in fact isomers, having the same overall formula, though different structures. In all the further reactions of the sequence, only glyceraldehyde-3-phosphate is used by the cell. So another enzyme now converts the dihydroxyacetone-phosphate into the isomeric glyceraldehyde-3-phosphate, thus making sure that both halves of the original hexose molecule are used up. The enzyme concerned is *triose phosphate isomerase*:

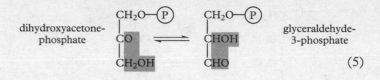

dihydroxyacetone-
phosphate

glyceraldehyde-
3-phosphate

(5)

The next stage in the degradation is the *oxidation* of glyceraldehyde-3-phosphate to 3-phosphoglyceric acid. In principle, one could write the reaction as

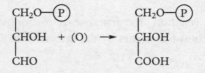

but in fact the details of the reactions are more complex.

Oxidation, we know, is an energy-yielding process, and the cell must make use of the energy released during oxidation of the glyceraldehyde-3-phosphate to make a molecule of ATP.

The enzyme concerned is a dehydrogenase, and for oxidation to proceed the coenzyme NAD is required. We have in this oxidation, then, the first example of the dehydrogenase reactions that we discussed in the previous chapter. To add complexity to the picture it was found that the enzyme is complex. It contains a sulphydryl group and in fact has three molecules of NAD bound to it – the enzyme also has a requirement for ADP and inorganic phosphate and during the series of reactions ATP is formed directly. This process is called *substrate level* phosphorylation to distinguish it from ATP synthesis coupled to the electron transport chain, which is oxidative phosphorylation.

The reaction occurs in three stages. In the first stage, the sulphydryl group of the enzyme and the NAD are reduced and glyceraldehyde-3-phosphate is phosphorylated by inorganic phosphate to 1,3,diphosphoglyceric acid:

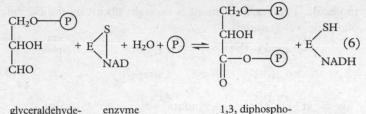

glyceraldehyde-   enzyme   1,3, diphospho-
3-phosphate   glyceric acid

ATP is now formed from ADP on a second site on the enzyme
by the reaction

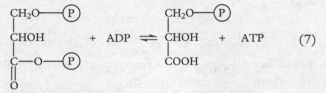

The enzyme must now be oxidized back to its original form, by
transferring hydrogens from the bound NAD to NAD free in
solution in the cell's cytoplasm. Thus the third stage of the reaction is

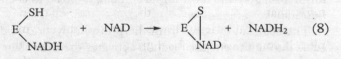

One molecule of glucose yielded two of glyceraldehyde-3-
phosphate, at the expense of two of ATP. Now, two molecules
of glyceraldehyde-3-phosphate have been oxidized to two of
phosphoglyceric acid, *making* two molecules of ATP in the
process. The net energy gain so far is thus zero, but we must
not forget that oxidation of glyceraldehyde-3-phosphate (reaction 6) *also* gave us two molecules of $NADH_2$. And we know
from the last chapter that these can also be oxidized by the
hydrogen carrier line, to give ATP in their turn.

Before the next phase of glycolysis can occur, the phosphoglyceric acid must be internally rearranged, by the transfer of
the remaining phosphate from position 3 to position 2 of the

167

molecule. The rearrangement is brought about by the enzyme *phosphoglyceromutase:*

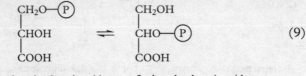

$$
\begin{array}{ccc}
CH_2O-\boxed{P} & & CH_2OH \\
| & & | \\
CHOH & \rightleftharpoons & CHO-\boxed{P} \\
| & & | \\
COOH & & COOH
\end{array}
\qquad (9)
$$

3-phosphoglyceric acid   2-phosphoglyceric acid

The tenth reaction of the series removes the elements of water from the phosphoglyceric acid, to yield phosphoenolpyruvic acid:

$$
\begin{array}{ccc}
CH_2OH & & CH_2 \\
| & & \| \\
CHO-\boxed{P} & \rightleftharpoons & C-O-\boxed{P} \quad + \quad H_2O \\
| & & | \\
COOH & & COOH
\end{array}
\qquad (10)
$$

2-phosphoglyceric acid   phosphoenolpyruvic acid

The responsible enzyme is *enolase*. Enolase requires Mg^{2+} ions as activators. Sodium fluoride, which traps magnesium by forming a complex compound with it, inhibits enolase; this is probably the most important single poisonous effect of fluoride.

The removal of water from the phosphoglyceric acid molecule during the enolase reaction activates the phosphate bond at carbon atom 2. Thus another synthesis of ATP becomes possible, catalysed by *pyruvic phosphokinase*:

$$
\begin{array}{ccc}
CH_2 & & CH_2 \\
\| & & \| \\
C-O-\boxed{P} \quad + \quad ADP \rightarrow & & C-OH \quad + \quad ATP \\
| & & | \\
COOH & & COOH
\end{array}
\qquad (11)
$$

phosphoenol-   (enol) pyruvic
pyruvic acid   acid

The product of this reaction, apart from ATP, is *(enol)pyruvic acid*, which changes spontaneously into the isomeric *pyruvic acid:*

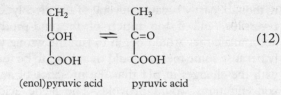

$$CH_2 \atop \| \atop COH \atop \| \atop COOH \quad \rightleftharpoons \quad CH_3 \atop \| \atop C=O \atop \| \atop COOH \qquad (12)$$

(enol)pyruvic acid   pyruvic acid

At this point we may stop for a moment and take stock of what has happened. Starting with one molecule of glucose, we used two molecules of ATP to phosphorylate it to fructose-1,6-diphosphate. We then split the fructose into two molecules of the 3-carbon glyceraldehyde phosphate. Each molecule of glyceraldehyde phosphate in turn yielded two of ATP, one at reaction (7), the other at reaction (11). Thus for each glucose two ATPs are spent and four synthesized, a net gain of two. But we have also converted two molecules of the coenzyme NAD into $NADH_2$ (reactions 5 and 6), which can be oxidized back to NAD, and in doing so produce another *six* molecules of ATP. Thus under aerobic conditions, another six ATPs are formed, and the net gain during glycolysis is eight ATPs.

But glycolysis can also operate, as we said at the beginning of this chapter, when *no* oxygen is present. Under these conditions there has to be some other method of reoxidizing the $NADH_2$. The way the animal cell does this is by using the $NADH_2$ to *reduce* pyruvic to lactic acid, thus:

$$CH_3 \atop \| \atop C=O \atop \| \atop COOH \quad + NADH_2 \rightleftharpoons \quad CH_3 \atop \| \atop CHOH \atop \| \atop COOH \quad + NAD \qquad (13)$$

pyruvic acid   lactic acid

the responsible enzyme being lactic dehydrogenase, an enzyme we have already met.

Reaction (13) accounts for the production of lactic acid which is observed during anaerobic glycolysis; the lactic acid is the end-product of the glycolysis reactions, and, unless oxygen is present, it cannot be metabolized further. The cell cannot afford to let large concentrations of lactic acid build up within it, though, partly because several of the glycolytic reactions are reversible, and if too much of the end-product began to accumulate they would begin to run the wrong way, and partly because some means would then have to be found of coping with the changes in pH that the presence of an acid in large concentrations would provoke. The lactic acid is therefore allowed to diffuse out of the cell and to be washed away in the bloodstream.

This completes the pathway known as glycolysis; the whole sequence is summarized in Figure 22.

GLUCOSE OXIDATION

The Embden–Meyerhof pathway ends with the production of pyruvic acid: when, however, oxygen is not in short supply, the pyruvic acid is itself further oxidized to carbon dioxide:

$$CH_3.CO.COOH \ + \ 5(O) \ \longrightarrow \ 3CO_2 \ + \ 2H_2O \qquad (14)$$

The full mechanism of this oxidation involves a whole series of reactions, and during it a considerable amount of energy is released, providing for as much as three-quarters of the cell's normal needs. Considerable experimental difficulties lay in the way of its complete unravelling, partly because a number of cofactors are involved, and partly because the process involves a *cycle* of reactants which are continually being broken down and reformed.

The cyclic mechanism is important because, as well as providing the cell with energy, glycolysis and glucose oxidation also perform another role as providers of the building blocks and intermediates for other pathways of metabolism. A cycle of metabolism enables these intermediates to be channelled to other reactions, occurring in other parts of the cell, without blocking the overall pathway by depletion. A cyclic pathway also makes it very easy for intermediates formed by the breakdown of substances other than glucose to enter and participate in the cycle, so that the cell can maximize the amount of energy it can extract from these breakdown products. However, it is

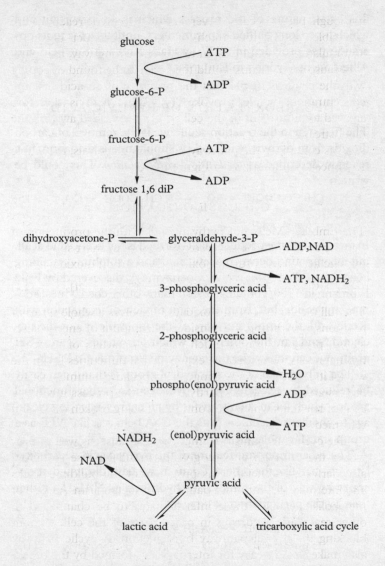

Figure 22. The glycolytic pathway

this cyclic nature of the process which is so important and which for so long baffled experimenters, until the clue to its operation was provided in 1937 by Hans Krebs at Sheffield and Albert Szent-Györgyi in Budapest.

DECARBOXYLATION OF PYRUVIC ACID

The first step in the reaction sequence is the removal of carbon dioxide from pyruvic acid, and its simultaneous oxidation, in a reaction described as *oxidative decarboxylation*. This could be written

$$CH_3.CO.COOH \ + \ (O) \ \longrightarrow \ CH_3.COOH \ + \ CO_2 \qquad (15)$$

pyruvic acid   acetic acid

In fact, though, the product that is formed is not acetic acid at all, but a compound between acetic acid and a sulphur-containing coenzyme called *coenzyme A*. Coenzyme A, discovered by Fritz Lipmann in 1950, contains, like so many other coenzymes, adenine nucleotide, but, from the point of view of its reaction with acetic acid, the important part of the molecule is not the nucleotide end but the other end which terminates in an —SH group. We can thus write the coenzyme simply as CoA.SH.

The importance of the sulphydryl group is that it can easily be reduced by exchanging its hydrogen for other functional groups which it can then transfer by an energy-releasing cleavage to another molecule – thus the CoA behaves, like ATP, as a group transfer molecule.

The other important feature of the pyruvic acid decarboxylation is that, like the other oxidations we have discussed, it operates by way of the removal of hydrogen as $NADH_2$, rather than by the participation of oxygen:

$$CH_3.CO.COOH \ + \ CoA.SH + \ NAD \ \longrightarrow$$

pyruvic acid $CH_3CO — SCoA + CO_2 + NADH_2$ (16)

acetyl-CoA.

In writing this reaction, we have not revealed the full complexity of the events surrounding pyruvate decarboxylation.

The reaction sequence actually involves three separate enzymes complexed together, and three cofactors. One of these cofactors is vitamin B_1, whose chemical name is thiamine pyrophosphate (TPP). The participation of thiamine in this reaction had been suspected ever since the 1920s, when Rudolph Peters at Cambridge found that in animals whose diet was lacking in vitamin B_1, pyruvic acid accumulated in the bloodstream, but it was only in the 1980s that the exact mechanism of the reaction has been worked out. The second cofactor is another sulphur-containing compound, lipoic acid, which is alternately oxidized and reduced during the reaction. The third cofactor is, of course, CoA.

In the currently accepted reaction scheme for equation (16) pyruvic acid first combines with the thiamine pyrophosphate to yield acetyl-TPP and carbon dioxide, a reaction catalysed by the first enzyme of the complex, pyruvic decarboxylase. The acetyl group is then transferred, first to lipoic acid by the enzyme lipoic acid reductase, releasing TPP, and then from lipoic acid to CoA.SH leaving the lipoic acid attached to its enzyme but in its reduced state, as dihydrolipoic acid. The third enzyme now oxidizes the dihydrolipoic acid to its original condition by effecting the transfer of its H_2 first to FAD, yielding $FADH_2$ and from thence to NAD, yielding $NADH_2$, thus completing the reaction sequence shown formally in equation (16) above. This complex sequence is probably required partly because the decarboxylation is strongly exergonic and hence must be coupled to a series of endergonic reactions in the synthesis of group transfer molecules if energy is not to be wasted. In addition, this is the first reaction we have come across in which the splitting of a carbon chain to yield the ultimate oxidative product, CO_2, occurs.

Citric acid cycle

By converting pyruvic acid to acetyl-CoA we have removed one CO_2 molecule and produced one molecule of reduced $NADH_2$. It still remains to oxidize the remaining two carbon atoms of the acetyl residue to carbon dioxide. It is here that we

Figure 23. The citric acid cycle

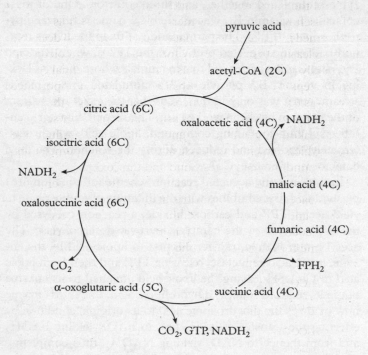

enter into the cycle of reactions that is associated with the name of Hans Krebs. The outline of the scheme is shown in Figure 23.

The essential feature of the Krebs cycle elucidated during the 1930s and 1940s is that the 2-*carbon* fragment of acetyl-CoA is made to combine with a 4-*carbon* acid, to yield the very reactive 6-*carbon* citric acid molecule. This reaction, necessitating as it does the formation of a carbon–carbon bond, is energy requiring – the energy, of course, comes from the splitting of the 2C fragment from its group transfer molecule CoA. Citric acid is then broken down again to give first the 5-*carbon* α-oxoglutaric acid and then the 4-*carbon* succinic acid, giving off two molecules of carbon dioxide in the process.

The succinic acid is then converted once more into oxaloacetic acid, and the entire cycle can start over again.

The useful thing about this merry-go-round as far as the cell is concerned is that it provides that critical feature of all desirable energy-releasing systems – a way for the acetic acid molecule to be carefully picked to pieces so as to release its potential energy in small steps. The cycle, as we shall find, is of great importance not only as the final stage of glucose oxidation, but in the metabolism of many other substances as well. We shall later see how, in the breakdown of both fatty and amino acids, products are formed which can enter the cycle at one of several points, to be whirled round the merry-go-round and thus oxidized.

In the first reaction of the cycle (17), acetyl-CoA combines with oxaloacetic acid to form citric acid. This reaction is *energy-requiring* and it is driven at the expense of acetyl-CoA. Reduced coenzyme A, CoA.SH, is released in the process, and the enzyme for the reaction is the *citrate condensing enzyme*.

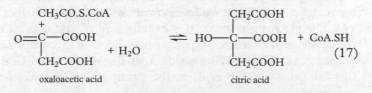

Citric acid is then converted to its isomer, isocitric acid, by way of the enzyme *aconitase*, which functions, in effect, by removing a molecule of water from the citric acid (producing aconitic acid, which, however, remains bound to the enzyme and is not released into the cytoplasm at all) and then replacing the water molecule isomerically (18):

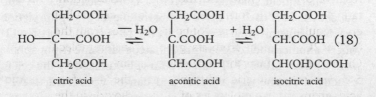

Isocitric acid is now *oxidized* to oxalosuccinic acid and then *decarboxylated* to give α-oxoglutaric acid. Again, both reactions are catalysed by the same single enzyme, *isocitric dehydrogenase*. In the process, a molecule of NAD is reduced to $NADH_2$. (There is also another form of isocitric dehydrogenase which uses the alternative coenzyme to NAD, NADP, and forms $NADPH_2$; the significance of this alternative enzyme will become clear in Chapter 8.) These reactions are shown in equation 19:

$$
\begin{array}{lll}
CH_2COOH & CH_2COOH & CH_2COOH \\
| & | & | \\
CH.COOH + (NAD) \rightarrow & CH.COOH + (NADH_2) \rightarrow & CH_2 + CO_2 \quad (19) \\
| & | & | \\
CHOH.COOH & CO.COOH & CO.COOH \\
\text{isocitric acid} & \text{oxalosuccinic acid} & \text{α-oxoglutaric acid}
\end{array}
$$

α-oxoglutaric acid resembles pyruvic acid in that it has the keto-group (CO) next to the acidic (COOH) group. It is thus capable of taking part in a similar reaction to the one described for pyruvic acid, that of *oxidative decarboxylation* (see equation 16). The mechanism involved is exactly similar, and demands the presence of thiamine pyrophosphate, lipoic acid, coenzyme A, and NAD. The resulting product, analogous to acetyl-CoA formed from pyruvic acid, is the group transfer molecule succinyl-CoA (20).

$$
\begin{array}{l}
CH_2COOH \\
| \\
CH_2 + CoA.SH + NAD \rightarrow \quad \begin{array}{l} CH_2.COOH \\ | \\ CH_2.CO.S.CoA + NADH_2 + CO_2 \end{array} \quad (20) \\
| \\
CO.COOH \\
\text{α-oxoglutaric acid} \qquad\qquad\qquad \text{succinyl-CoA}
\end{array}
$$

The subsequent splitting of succinyl-CoA releasing CoA and succinic acid is used to drive a substrate level phosphorylation reaction, but this time it results in the formation, not of ATP, but of another nucleoside triphosphate, guanosine triphosphate (GTP) from GDP. GTP can of course be utilized by the cell in exactly the same way as ATP, and can be converted to ATP directly at the expense of ADP, as shown below.

$$\begin{array}{ccc}
\underset{|}{CH_2COOH} & & \underset{|}{CH_2COOH} \\
CH_2CO.S.CoA + GDP + \textcircled{P}i & \rightleftharpoons & CH_2COOH + GTP + CoA.SH
\end{array} \qquad (21)$$

$$GTP + ADP \rightleftharpoons GDP + ATP \qquad (22)$$

Finally, succinic acid is converted to oxaloacetic acid by way of fumaric and malic acids, the reactions being catalysed by *succinic dehydrogenase*, *fumarase* and *malic dehydrogenase*. One molecule of reduced flavoprotein is formed at the succinic dehydrogenase catalysed step, one of $NADH_2$ by malic dehydrogenase (23):

$$\begin{array}{ccc}
\underset{|}{CH_2COOH} & \overset{-2H}{\rightleftharpoons} & \underset{||}{CH.COOH} \\
CH_2COOH & & CH.COOH \\
\text{succinic acid} & & \text{fumaric acid}
\end{array}$$

$$\begin{array}{ccccc}
& \overset{+ H_2O}{\rightleftharpoons} & \underset{|}{CHOH.COOH} & \overset{-2H}{\rightleftharpoons} & \underset{|}{CO.COOH} \\
& & CH_2COOH & & CH_2COOH \\
& & \text{malic acid} & & \text{oxaloacetic acid}
\end{array} \qquad (23)$$

and, with the re-formation of oxaloacetic acid, the cycle is completed (see Figure 23).

It may be observed that, for the cycle to revolve, a supply of oxaloacetic acid is required with which the acetyl-CoA can react. Of course, oxaloacetic acid is being constantly re-formed as each revolution of the cycle is completed, but, nonetheless, it is not surprising to find that many cells, especially in bacteria, take precautions against running out of so essential a 'primer' by having an alternative source of oxaloacetic acid. The alternative source is provided by an enzyme, *oxaloacetic decarboxylase*, which catalyses the direct 'fixing' of carbon dioxide on to pyruvic acid to form oxaloacetic acid as in equation (24):

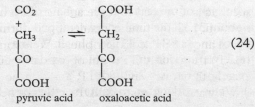

$$\begin{array}{ccc}
CO_2 & & COOH \\
+ & & | \\
\underset{|}{CH_3} & \rightleftharpoons & \underset{|}{CH_2} \\
\underset{|}{CO} & & \underset{|}{CO} \\
COOH & & COOH \\
\text{pyruvic acid} & & \text{oxaloacetic acid}
\end{array} \qquad (24)$$

177

By a masterly piece of economy, then, many cells have a mechanism whereby the primer with which the acetyl-CoA derived from pyruvic acid must combine in order for oxidation to occur is *itself* a product of a reaction deriving from pyruvic acid.

ATP PRODUCTION

We may now try to arrive at the figure for total ATP production during the cycle. We have:

(a) pyruvic ⟶ acetyl-CoA + 1 mol.$NADH_2$ = 3 ATP
(b) isocitric ⟶ α-oxoglutaric + 1 mol.$NADH_2$ = 3 ATP
(c) α-oxoglutaric ⟶ succinyl-CoA + 1 mol.$NADH_2$ = 3 ATP
(d) succinyl-CoA ⟶ succinic   = 1 GTP *
(e) succinic ⟶ fumaric + 1 mol.FPH_2 = 2 ATP
(f) malic ⟶ oxaloacetic + 1 mol.$NADH_2$ = 3 ATP

Total   15 ATP

(*1 GTP = 1 ATP)

Thus one molecule of pyruvic acid provides fifteen molecules of ATP.

As each molecule of glucose gives rise to *two* molecules of pyruvic acid by glycolysis, oxidation of glucose by way of the citric acid cycle yields thirty molecules of ATP. To this must be added the eight already produced between glucose and pyruvic acid and we conclude that the complete oxidation of glucose to carbon dioxide and water yields no less than thirty-eight molecules of ATP in all. As the energy released during the hydrolysis of the terminal phosphate of each molecule of ATP is about 31 kJ, the total energy trapped during the oxidation of glucose is 38×31 or 1178 kJ in all. We may compare this with the equation for the complete oxidation of glucose, which (page 108) releases 2820 kJ.

When working at maximum efficiency, therefore, the cell

manages to trap and retain in usable form no less than 42 per cent of the energy released during the burning of its fuel. This efficiency is equal to that of many oil- or coal-fired power stations. But we must bear in mind that, like the power station, this calculation applies only to the conversion of energy in the boiler itself. The moment we come to calculate the *overall* efficiency, we have to take into account the losses of energy that occur in the transport of fuel *to* the boiler, and in the transport of energy *away* from the boiler subsequently. The overall efficiency of power-houses, or of the cell, is much lower than the figure of 42 per cent might suggest.

OTHER SYSTEMS OF GLUCOSE OXIDATION

The Embden–Meyerhof pathway and the citric acid cycle, whilst amongst the most universal of glucose oxidation systems found in nature, are by no means the only ones. Many variants on the citric acid cycle are known to exist in different organs and species, and several alternative cycles or 'shunts' have been suggested. That the cycle itself occurs in almost all tissues has been confirmed abundantly, in particular by the use of isotopic tracers, but it is often difficult to obtain a quantitative estimate of the extent to which the glucose oxidized in the living cell uses the cycle or is diverted to other routes.

One of the most common of the alternative pathways, which may contribute anything up to 30–40 per cent of the glucose oxidized in some tissues, is the so-called Warburg–Dickens pathway, again, in its basic outlines, the work (during the 1930s) of the biochemists after whom it is named. (Another name for the same route which is sometimes used is 'the pentose phosphate shunt'.) Like the Embden–Meyerhof route, it starts with a glucose-6-phosphate, but oxidizes it and decarboxylates it in an NADP-linked enzyme system to the 5-carbon sugar phosphate *ribulose-5-phosphate*, an isomer of ribose. Six of these ribulose-5-phosphate molecules, formed from six molecules of glucose-6-phosphate, are then combined and shuffled through an intricate maze of reactions which result in the resynthesis of five molecules of the glucose phosphate once more.

The net result is that the equivalent of one molecule of glucose is oxidized to carbon dioxide and in the process thirty-six molecules of ATP are formed – two less than are given in the Embden–Meyerhof pathway coupled to the citric acid cycle. The advantage of the shunt system is that fewer enzymes are required, as there are fewer reactions involved, and also that the reactions themselves, and their enzymes, are very similar to those utilized by plants during photosynthesis.

In addition, the pathway provides pentose sugars which are precursors for nucleic acid synthesis and a source of the reduced coenzyme $NADPH_2$. This substance is used, as we shall see, in the synthesis of fatty acids, steroids, and amino acids, and cannot be replaced by $NADH_2$ in these reactions.

It has been suggested that the shunt system evolved prior to the Embden–Meyerhof pathway, possibly before plants and animals had properly distinguished themselves from their simple ancestors (see Chapter 14). If this were so, the Embden–Meyerhof route would represent a later, more sophisticated, development.

GLYCOGEN BREAKDOWN

A final point should be made about the entry of glucose into the oxidative systems. We have up till now regarded the glucose as entering the cell as 'free' sugar, and as being phosphorylated by hexokinase prior to oxidation. In those cells which contain their own supplies of glycogen – mainly liver and muscle – glucose is also provided by the breakdown of glycogen. An enzyme called *phosphorylase* ruptures the linkages that bind the glucose units together in the glycogen chain. The reaction requires inorganic phosphate, and glucose units are released as glucose-1-phosphate (25):

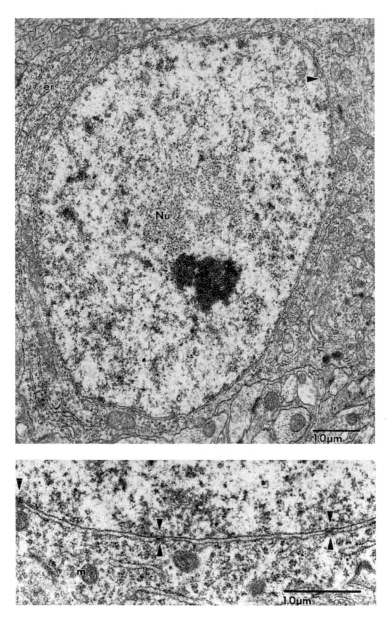

1 NUCLEUS (*above*) **Nu** – nucleus; **er** – endoplasmic reticulum; arrow shows pore in nuclear membrane. NUCLEAR MEMBRANE (*below*). Arrows show nuclear pores.

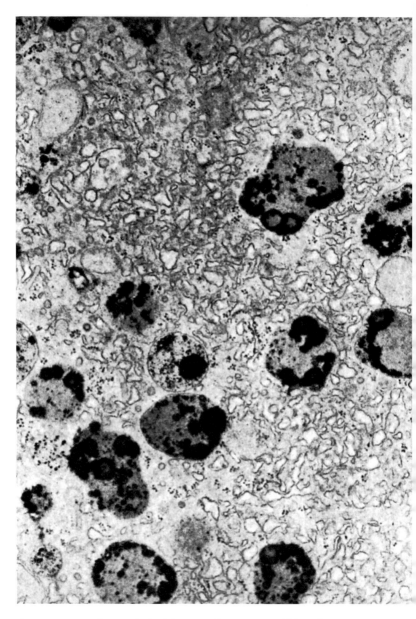

2 LYSOSOMES. Large dark objects are lysosomes within a liver cell.

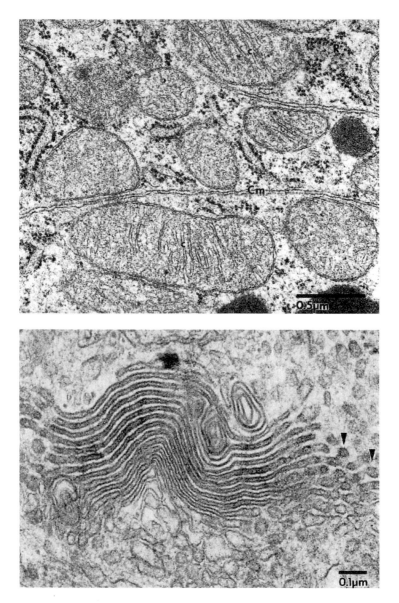

3 MITOCHONDRIA (*above*). **C** – mitochondrial cristae; **r** – ribosomes. GOLGI APPARATUS (*below*). Arrows show blebs.

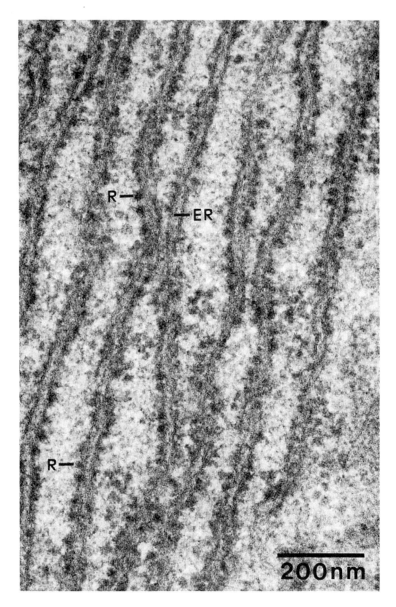

4 ENDOPLASMIC RETICULUM. Cytoplasm packed with endoplasmic reticulum (**ER**) and individual ribosomes (**R**).

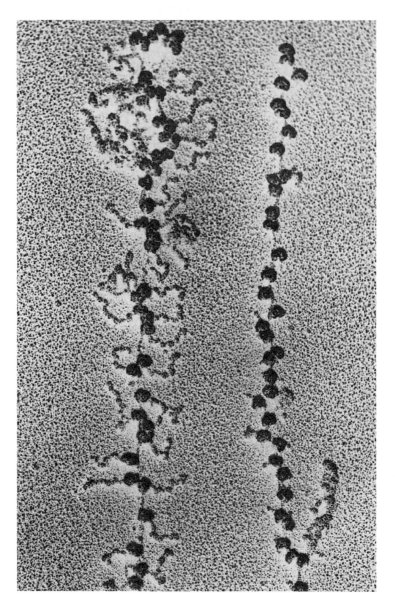

5 POLYSOME. Two fragments of one polysome – 3′ end mRNA and
growing protein chains on 5′ end (*left*).

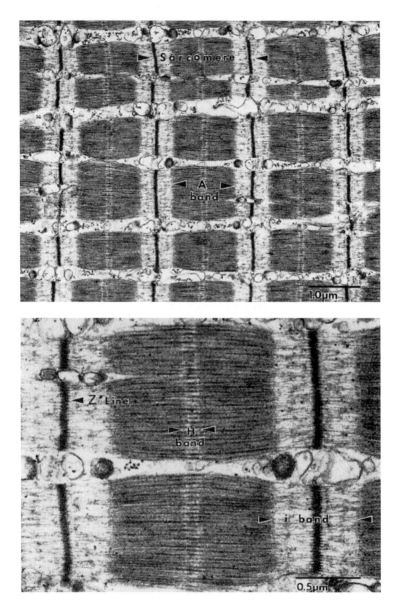

6 MUSCLE. (*Above*) Striated (voluntary) muscle, showing pattern of A, I, Z, and H bands which compose contractile material. (*Below*) At higher power, one complete band showing thick and thin filaments.

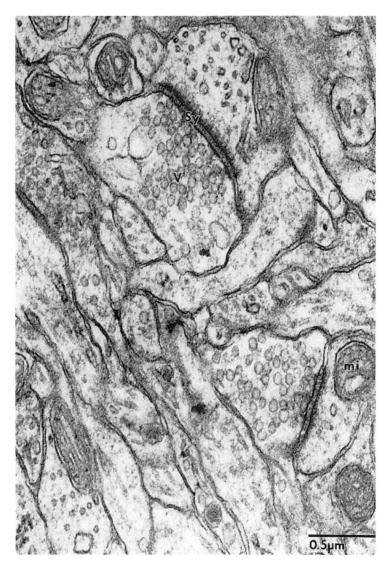

7 SYNAPSE (**Sy**). Junction between one nerve cell and a second is
made at darker, ticker region of membranes separating them. The
presynaptic cell contains mitochondria (**mi**) and many, much smaller, circular synaptic vesicles (**V**), containing transmitter substance;
postsynaptic side has no such vesicles.

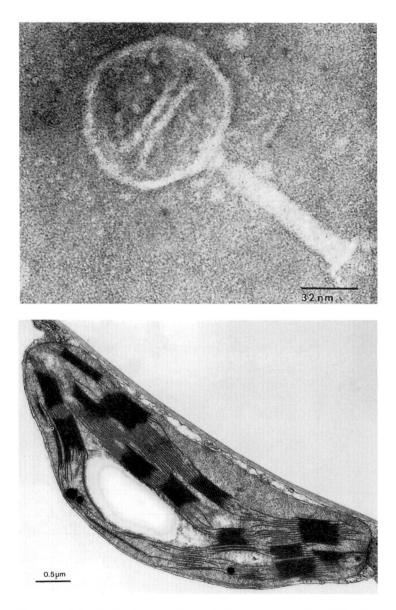

8 VIRUS (*above*). Single virus (bacteriophage) – a nucleic acid head
surrounded by a protein jacket and tail. CHLOROPLAST (*below*).
From leaf of *Impatiens* – 'Busy Lizzie'.

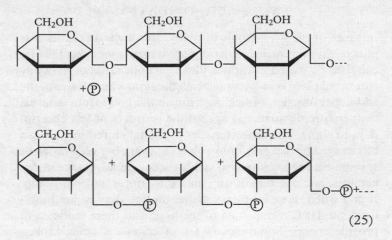

$$(25)$$

The glucose-1-phosphate is then converted by the isomerizing enzyme *phosphoglucomutase* into glucose-6-phosphate, which is in the direct path of glucose metabolism we have already discussed. We can thus draw the first stages of glucose metabolism as in Figure 24.

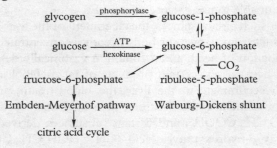

Figure 24

At the entry to glucose metabolism, then, stand two enzymes, hexokinase and phosphorylase. We shall later (Chapter 11) see how, because of this strategic placing at the starting-gates to glucose catabolism, these enzymes are able to perform a critical role in *controlling* the overall rate of glucose oxidation and of regulating it to the demands of the cell for energy.

FATTY ACID OXIDATION

In times of plenty, animals turn any carbohydrate that is surplus to their immediate energy requirements into fat. This fat can then be stored in various inconspicuous areas of the body (the usual place is in layers beneath the skin, where it serves the additional function of helping insulate the body from external temperature fluctuations) against the prospect of less cheerful days to come. The characteristics of the fat stored in cells can be changed by diet and, unlike glycogen, vary between different species. Fat metabolism also yields water – an advantage in dry climates, and one reason why the camel stores fat in its hump. If and when these lean days arrive, the fat 'depots' are mobilized, the fat is transported to the liver, and there oxidized to provide energy. Some dietary fat, of course, is being broken down the whole time even in the normal, healthy animal, but it only really comes into its own as a major contributor to the overall energy-balance of the body when the animal is hard-pressed for carbohydrate, due to starvation, or when, as in diabetes, the metabolic pattern of the body becomes disturbed, so that it is unable to use glucose effectively.

Under such circumstances, the fat content of the liver begins to rise as fat from all over the body becomes concentrated there in anticipation of need, and a condition graphically described as 'fatty liver' arises; certain products of fat metabolism also begin to accumulate in the liver, the bloodstream, and the urine; these include the 4-carbon keto acid acetoacetic acid ($CH_3CO.CH_2.COOH$), and certain substances derived from it, notably β-hydroxybutyric acid ($CH_3CHOH.CH_2.COOH$) and acetone ($CH_3CO.CH_3$). These substances go under the rather medieval sounding description of 'ketone bodies', and it is their presence which represents one of the greatest dangers in diabetes, for they are poisonous.

Fat metabolism, then, has features which are of considerable medical interest, and it is perhaps for this reason that its study stretches back as far as that of glucose oxidation. Knowledge of the mechanics of the two processes has grown in parallel, the result of the work of a distinguished line of investigators from the 1920s until the 1990s. We need not try to trace the many turns this research has taken, but instead can content ourselves with setting out current thinking on it.

Fats proper consist of long-chain fatty acids in ester linkage with glycerol (page 78), and the first step in their oxidation is their hydrolysis by *lipase* to release glycerol and the free fatty acids. The glycerol can be phosphorylated by ATP to give glycerol phosphate, and then oxidized by an NAD-linked enzyme to dihydroxyacetone phosphate, which, it will be recalled (page 166), lies on the direct pathway of glucose degradation in the Embden–Meyerhof scheme. From here on, glycerol follows the routes we have already mapped out to carbon dioxide and water (26):

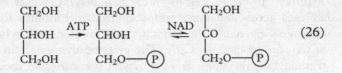

The problem of fat oxidation then resolves itself into that of the fate of the fatty acids derived from the fat. These, the straight chain saturated acids stearic (18-carbon) and palmitic (16-carbon), are oxidized, according to experimental results obtained by Knoop early in the twentieth century, by a process which splits off the carbon atoms two at a time; thus stearic is converted first to palmitic, then to a 14-carbon acid, then a 12-carbon acid, then a 10-carbon, and so on, until the molecule is whittled right down to the 4-carbon acetoacetic acid, which is split in its turn into two 2-carbon acetic acid fragments. The removal of carbon atoms two at a time like this is called β-oxidation, as it occurs by the oxidation of the carbon atoms two places away from (in chemical parlance, in the β-position to) the acidic group at the end of the fatty acid chain. Unsaturated fatty acids follow a slightly different route.

In 1949 Eugene Kennedy and Albert Lehninger in New

York showed that fatty acids are oxidized in the mitochondria. It is now known that they are activated in the mitochondrial membrane before they enter the matrix where they are oxidized. Energy from ATP is used to form a thioester linkage between the carboxyl group of a fatty acid and the sulphydryl group of CoA. The enzyme responsible is *acyl CoA synthetase* (also called *fatty acid thiokinase).*

$$R.CH_2.CH_2.CH_2.COOH + CoASH + ATP \longrightarrow$$
$$R.CH_2.CH_2.CH_2.CO.S.COA + AMP + 2\,\textcircled{P}_i \qquad (27)$$

Because long-chain acyl CoA molecules cannot easily cross the mitochondrial membrane a special transport mechanism is needed. *Carnitine* carries the activated long-chain fatty acids across the inner mitochondrial membrane. The fatty acyl group is transferred from the sulphur atom of CoA to the hydroxyl group of carnitine to form acyl carnitine, which diffuses across the inner mitochondrial membrane. When it has safely arrived in the mitochondrial matrix, the acyl group is transferred back to CoA. The enzyme that catalyses these transacylase reactions is *fatty acyl CoA: carnitine fatty acid transferase.*

Once inside the mitochondrion, acyl CoA is oxidized by a flavoprotein dehydrogenase enzyme.

$$R.CH_2.CH_2.CH_2.CO.SCoA + FP \longrightarrow$$
$$R.CH_2.CH = CH.CO.SCoA + FPH_2 \qquad (28)$$

This reaction produces a double bond, and the next step is hydration across this double bond by a *lyase* to yield β-hydroxyacyl CoA.

$$R.CH_2CH = CH.CO.CoA + H_2O$$
$$\rightleftharpoons R.CH_2\underset{\overset{|}{OH}}{CH}.CH_2CO.SCoA \qquad (29)$$

Another dehydrogenase, this time NAD-linked, converts the hydroxy- to the β-ketoacyl-CoA (30):

$$R.CH_2CH.CH_2CO.SCoA \quad + \quad NAD$$
$$\overset{|}{\underset{OH}{}}$$
$$\rightleftharpoons \quad R.CH_2C.CH_2CO.SCoA \quad + \quad NADH_2 \quad (30)$$
$$\overset{\|}{\underset{O}{}}$$

This keto acid is enabled to react with another molecule of coenzyme A (31) by *β-ketothiolase*:

$$R.CH_2C.CH_2.CO.SCo.A \quad + \quad CoASH \quad \rightleftharpoons \quad R.CH_2CO.SCoA$$
$$\overset{\|}{\underset{O}{}}$$
$$+ \quad CH_3.CO.SCoA \quad (31)$$

This last reaction succeeds in *splitting* the carbon chain of the acid, releasing acetyl-CoA and the CoA derivative of the fatty acid with *two fewer* carbon atoms than the original molecule of equation (27). This reaction is known as *thiolysis*. In it, a molecule of CoA is added to the fatty acyl residue without the need of energy from ATP. The thiolysis reaction itself provides enough energy for the addition to be made without need of an external energy source as well. As well as releasing a molecule of acetyl-CoA, the thiolysis reaction therefore also produces a fatty acyl-CoA which can now take part in the same reaction sequence that we have already traced, beginning with reaction (28) above. That is, each time that a fatty acyl-CoA molecule passes through the sequence of reactions (28) to (31), of oxidation, hydration, oxidation, and thiolysis. it produces a daughter molecule of acetyl-CoA and an acyl-CoA molecule two carbons shorter than its parent but ready in its turn to breed in the same manner.

The fatty acid molecule thus undergoes a repetitive series of four reactions which steadily chip away at its carbon skeleton, releasing the fragments as acetyl-CoA. We can show the complete process for the 6-carbon *hexanoic acid* like this:

$$CH_3CH_2CH_2CH_2CH_2COSCoA$$

$$\downarrow$$

$$CH_3CH_2CH_2CO.SCoA + CH_3CO.SCoA$$
$$+ \quad NADH_2 \quad + \quad FPH_2 \qquad (32)$$

$$\downarrow$$

$$CH_3CO.SCoA \quad + \quad CH_3CO.SCoA$$
$$+ \quad NADH_2 \quad + \quad FPH_2$$

and it then appears that from hexanoic acid we obtain three molecules of acetyl-CoA, two of $NADH_2$, and two of FPH_2. When the acetyl-CoA is oxidized through the citric acid cycle, each molecule gives twelve of ATP; the two molecules of $NADH_2$ are worth six ATPs, and the two flavoproteins another four. The yield of ATP produced by the complete biological oxidation of hexanoic acid is thus thirty-six plus six plus four or forty-six ATPs in all; as two are expended in forming hexanoyl-CoA from the hexanoic acid we began with, the net gain in ATP is forty-four molecules. This may be compared with the thirty-eight produced during the oxidation of glucose. The 6-carbon fatty acid thus releases *more* usable free energy than the 6-carbon carbohydrate.

However, the oxidation is probably not as efficient as it looks at first sight, for we have assumed in the calculation that all the acid is broken down into acetyl-CoA which is subsequently oxidized. But in practice, we have already noted that under conditions where fatty acids form the major energy source of the body, during starvation or diabetes, large amounts of the substance acetoacetic acid accumulate. Now one source of acetoacetic acid is the combination of two of the molecules of acetyl-CoA formed during fatty acid oxidation (33):

$$CH_3CO.SCoA \quad + \quad CH_3CO.SCoA$$
$$\rightleftharpoons CH_3COCH_2CO.SCoA + CoASH \qquad (33)$$

Evidence from isotopic tracer experiments suggests that acetoacetic acid is in fact formed in this way during fatty acid oxidation.

However, every molecule of acetoacetic acid or its derivatives that the liver accumulates and subsequently excretes means that two potential acetyl-CoA molecules have not been exploited so as to release their potential energy through oxidation by way of the citric acid cycle. Why? Particularly as fatty acid oxidation only occurs under conditions of carbohydrate starvation, one may well be surprised at the fact that the cell seems to squander a valuable source of energy in this wasteful way. To discard acetoacetic acid like this is to throw away fuel before it is more than half-burned.

The reason seems to be tied up with the way the citric acid cycle works. We mentioned earlier that, in order for the cycle to turn, acetyl-CoA must combine with oxaloacetic acid to form citric acid, and that, although in theory only catalytic amounts of oxaloacetic acid are required, since it is constantly re-formed as the cycle comes full circle, nonetheless the cell in practice seems to find it necessary to provide an additional source of oxaloacetic acid by way, for instance, of reaction (24) of page 177. In such reactions, oxaloacetic acid is produced from pyruvic acid. But pyruvic acid lies only on the pathway of glucose oxidation, and not on that of fatty acid oxidation. There is no pyruvic acid at all formed during the series of reactions we have just described. It follows that the supplies of oxaloacetic acid must be sharply diminished during carbohydrate starvation. Thus the citric acid cycle must gradually slow down, as a queue of acetyl-CoA molecules is formed, lining up and waiting to be linked to oxaloacetic acid and taken for a ride on the merry-go-round.

And as the amount of acetyl-CoA increases, an increasing number of molecules will tend to combine with one another by reaction (33). It seems then that, in the absence of carbohydrate, the cell, forced to turn for free energy to its fat supplies, finds that it cannot even use these as efficiently as would be possible were carbohydrate present to provide the essential oxaloacetic acid. This seems to be a strange weakness and inefficiency, and serves to emphasize once again that, in the animal world so far as energy is concerned, there is no substitute for sugar, even though the full oxidation of fats would yield more ATP than would the equivalent amount of glycogen.

OTHER SOURCES OF ENERGY

Fats, we have seen, are used as an energy-source only when the animal is starved of glucose. If starvation is prolonged, though, the body's stocks of fat begin to run down and finally are exhausted. It becomes necessary to search elsewhere for oxidizable substrate. The only remaining substances present in the body which can provide this last resort are the proteins. Proteins are so essential to survival that it is only with the greatest reluctance that the cell will turn to them as an energy source. Yet, if all other supplies fail, ultimately they too begin to be consumed to stave off the day when no more life-preserving ATP can be produced. It is as if, having burned all other fuel, one began in desperation to chop up the furniture and throw it too on the fire.

Obviously, the process cannot be continued indefinitely; under normal circumstances, the body maintains a 'nitrogen balance'; the amount of nitrogen taken in from the intestines as amino acids being exactly equivalent to the amount released during the breakdown and disposal of 'worn-out' protein and excreted as urea in the urine. As starvation proceeds, the nitrogen balance is tipped – increasing amounts of nitrogen are excreted as urea as the proteins of which it formed part are oxidized for energy, and the body begins to break down its protein more rapidly than it replaces it. The amounts of nitrogen excreted rise day by day and finally shoot up dramatically, as if the body despaired of being able to conserve any protein at all, and this last great rise is rapidly followed by death.

Apart from an understandable reluctance to waste good protein as an energy source, its use by the cell is complicated by the difficulties attached to providing for the safe disposal of the nitrogen contained in the protein. Useless to animals as a source of energy, the nitrogen tends to accumulate as ammonia, NH_3. But ammonia is dangerously poisonous; even very small amounts injected into the bloodstream can be lethal. The body is therefore committed to finding a non-toxic disposal system for the ammonia which protein metabolism produces.

This it does by converting the ammonia into the harmless urea, $CO(NH_2)_2$, which can be passed through the bloodstream to the kidneys and thence excreted into the urine without fear of self-poisoning.

The fact that the urine contains a fair amount of urea even while the body is healthy and functioning normally indicates that protein breakdown is occurring all the time; we have already seen how the proteins, along with other giant molecules which the cell produces, are constantly being destroyed and re-formed in the ceaseless self-renewal which the cell undertakes and which is described as the 'dynamic state of body constituents'. Obviously, the cell regains what energy it can during this breakdown process, and so, like the fats, amino acids and proteins contribute to the energy-balance of the body even when carbohydrate is plentiful. Under normal circumstances, though, their services to this cause must be very slight.

The conversion of protein nitrogen into urea is a process that takes place almost exclusively in the liver. (Dogs whose livers have been removed can live for some time on a nitrogen-free diet, but quickly die if fed protein as they now have no means of preventing nitrogen being released as ammonia and poisoning them.) The mechanism of urea formation was largely discovered by Hans Krebs in the 1930s; like that of acetyl-CoA oxidation, it is a *cyclical* process of considerable theoretical interest. Indeed it was Krebs's discovery of the urea cycle, pre-dating his work on the better-known tricarboxylic acid cycle that bears his name, which gave him the idea that cycles may be a general mechanism of metabolism (and, as it happens, also a Nobel Prize). But as it is not a source of energy to the cell, we shall not describe the urea cycle further here.

Such energy as the proteins can provide comes through the oxidation of the carbon skeletons of their constituent amino acids. As there are many amino acids, with quite widely differing structures, it is hardly surprising that there are several different oxidative routes available. In all of them, though, the principle of the degradation is similar; the amino acid is disembarrassed of its amino-nitrogen group at an early stage and the denuded molecule is converted into substances which lie on the direct pathway of either glucose or fatty acid oxidation, where it can be disposed of by reaction pathways we have already described. Amino acids which give rise to glucose oxidation and citric acid cycle intermediates are known as *glucogenic*, whilst those which are converted into substances lying on the pathway of fatty acid metabolism are *ketogenic*.

There are several ways of ridding the amino acid of its nitrogen. One such would be direct oxidation, in which the acid is first dehydrogenated, passing two hydrogens to an acceptor, and then hydrolysed by water, releasing ammonia (34, 35):

$$
\begin{array}{ccc}
\text{R.CHCOOH} & & \text{R.C.COOH} \quad + \quad 2\text{H} \\
| & \rightleftharpoons & \| \\
\text{NH}_2 & & \text{NH} \\
\text{amino acid} & & \text{imino acid}
\end{array}
\tag{34}
$$

$$
\begin{array}{ccc}
\text{R.C.COOH} \quad + \quad \text{H}_2\text{O} & & \text{R.CO.COOH} \quad + \quad \text{NH}_3 \\
\| & \rightleftharpoons & \\
\text{NH} & & \\
\text{imino acid} & & \text{keto acid}
\end{array}
\tag{35}
$$

The other product of reaction (35) is a keto acid, whose further fate depends on the exact nature of the R-group in it. Reactions of the type of (34) are catalysed by *amino acid oxidases*. The second reaction (35) is spontaneous and requires no enzyme. The most important of the amino acid oxidases is *glutamic dehydrogenase*, which converts glutamic acid into α-oxoglutaric acid (generating NADH$_2$ in the process) (36):

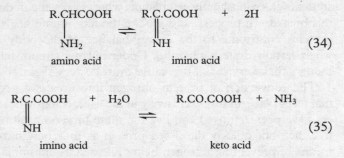

$$
\begin{array}{ccccc}
\text{COOH} & & \text{COOH} & & \\
| & & | & & \\
\text{CH}_2 & +\text{H}_2\text{O} & \text{CH}_2 & & \\
| & -2\text{H} & | & & \\
\text{CH}_2 & \rightleftharpoons & \text{CH}_2 & + & \text{NH}_3 \\
| & & | & & \\
\text{CHNH}_2 & & \text{CO} & & \\
| & & | & & \\
\text{COOH} & & \text{COOH} & & \\
\text{glutamic acid} & & \text{α-oxoglutaric acid} & &
\end{array}
\tag{36}
$$

The α-oxoglutaric acid is of course a member of the citric acid cycle series of acids, and we have already described its fate.

The importance, though, of the glutamic dehydrogenase reaction is that, apart from this enzyme, the only amino acid oxidases present in the cell are either rather weak enzymes, or else are ones that are specific not for the naturally occurring (–)isomers of the amino acids, but for the unnatural (+)isomers. In general, amino acids other than glutamic have a slightly different method of freeing themselves of nitrogen. This alternative route is called *transamination*, and transaminase enzymes exist for many amino acids so that, starting with any other amino acid and α-oxoglutaric acid, the exchange reaction (37) can occur:

$$
\begin{array}{ccccccc}
R & & COOH & & R & & COOH \\
| & & | & & | & & | \\
CHNH_2 & + & CH_2 & & CO & + & CH_2 \\
| & & | & & | & & | \\
COOH & & CH_2 & \rightleftharpoons & COOH & & CH_2 \qquad (37)\\
 & & | & & & & | \\
 & & CO & & & & CHNH_2 \\
 & & | & & & & | \\
 & & COOH & & & & COOH
\end{array}
$$

amino acid   α-oxoglutaric acid   α-keto acid   glutamic acid

Under the influence of transaminase, the amino acid swaps its NH_2 amino group for the keto group of α-oxoglutaric acid, releasing the appropriate keto acid and glutamic acid. The keto acid can go on its own metabolic way by way of reactions we have already seen in the glycolytic and glucose oxidation pathways, and glutamic acid can be deaminated by glutamic dehydrogenase to produce α-oxoglutaric acid once more. By this means reactions (36) and (37) can be coupled to provide yet another of the now familiar cyclical processes in which the cell delights (Figure 25). This way, virtually all amino acids are sucked into the metabolic whirlpool and made to yield their potential energy.

Figure 25

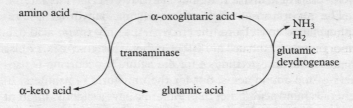

INTERCONVERSIONS OF FATS, AMINO ACIDS,
AND CARBOHYDRATES

All the cells of the body metabolize glucose. But fatty acids and
amino acids are utilized mainly in the liver, although some
amino acid metabolism can also occur in other tissues, such as
brain or kidney. When starvation conditions arise, although the
liver can get all the energy it needs by burning fat, other tissues
would tend to go short if there were no means of dispatching
energy to them. The most convenient way of sending packets
of energy over long distances in the body is to parcel it up as
glucose. So the liver needs to be able to convert the foodstuffs
it uses into sugar for the benefit of less versatile tissues. What
this demands is that the liver cells provide a system of revers-
ing the sequence of reactions that leads from glucose to acetyl-
CoA, for both fats and amino acids produce this substance
during their oxidation.

Fortunately, nearly all of the reactions that we have traced
out in describing the pathway of glucose oxidation are
reversible. Given the right conditions, all the enzymes which
help to break down fructose-1,6-diphosphate to pyruvic acid
(reactions 4 to 12) can be made to retrace their steps.
Fructose-1, 6-diphosphate itself is made from glucose with
the help of two molecules of ATP and the enzymes hexokinase, phosphofructokinase, and phosphohexoisomerase
(reactions 1 to 3, pages 164–5). The two ATP molecules
which are spent in 'priming' the hexose prior to breaking it

down are irrevocably lost, for both the hexokinase and phosphofructokinase reactions are irreversible. However, there exist phosphatase enzymes which can hydrolyse hexose phosphates to release the free hexose and, not ATP, but inorganic phosphate. With their help, fructose diphosphate may be reconverted into glucose:

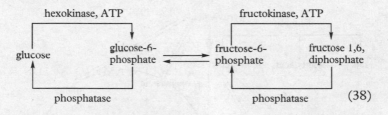

(38)

Thus pyruvic acid and glucose are connected by a reversible sequence of reactions. But fats and amino acids yield either acetyl-CoA or citric acid cycle intermediates like α-oxoglutaric acid, not pyruvic acid. In the glucose oxidation scheme, pyruvic acid is oxidatively decarboxylated to acetyl-CoA (equations 15 and 16, page 172) by a complex reaction sequence that appears to be virtually irreversible. In order to resynthesize pyruvic acid, we need to make use of a more roundabout method. The reactions on page 174 showed one way in which, starting with oxaloacetic (or malic) acid, some types of cell can resynthesize pyruvic acid. As both these substances are formed during the rotation of the citric acid cycle, both acetyl-CoA and the cycle intermediates can give rise to pyruvic acid. Although the details of the exact ways in which these reactions actually occur are by no means yet fully mapped out, and it is not certain whether all cells of all species contain this mechanism, it seems clear that there *are* ways in which the cell can, at a pinch, regain glucose at the expense of other foodstuffs. Equally, glucose itself can be used as the starting substrate for the synthesis of fatty acids, amino acids, fats, and proteins that is discussed in the next chapter. We can summarize all these interrelationships in one comprehensive diagram, Figure 26.

Figure 26. Interrelations of fats, carbohydrates, and proteins

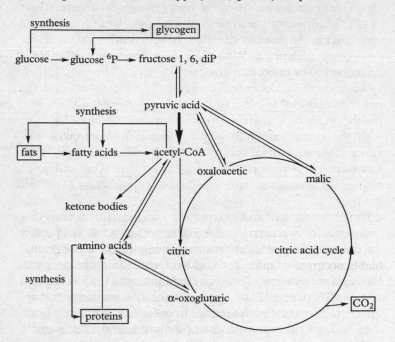

CHAPTER 9

The Synthetic Pathways

It is notoriously easier to destroy than to build. The methods by which the cell breaks down its constituents are indeed ingenious, but their prime purpose is simply to provide systems for pulling molecules apart and releasing their potential energy with the minimum of fuss and bother. The molecules are knocked about, pushed and pulled into convenient shapes, like limousines in a breaker's yard, with scant ceremony and little respect for what they once were, until at last all that is left of them can be pounded out through the citric acid cycle and dumped as carbon dioxide.

Synthesis, by contrast, is a complex and meticulous job, like the assembly-line of a car factory, turning out precision-built, highly accurately finished models with the minimum of flaws. Biosynthetic pathways demand that we start with the most primitive of the cell's building blocks – simple organic acids, carbon dioxide, and a nitrogen-source – and from them build a galaxy of macromolecules, carbohydrates, lipids, proteins, nucleic acids.

By contrast with some microorganisms or with plants, animals are not very biosynthetically versatile. Plants and some bacteria can build carbon chains starting only with carbon dioxide and the sun's energy, whilst many other bacteria can grow on such unnutritive-seeming substances as acetic acid. The animal cell needs much more; before it can set to work it must be supplied with sugar, several different sorts of amino acid, and a variety of vitamins. In truth, the animal lives parasitically on the ability of others to synthesize the substances it needs; freed from the necessity of staying in one place with its arms out to catch the sun's rays, or of touring the countryside for a supply of acetic acid, it can instead do other, more interesting things. Even so, one mustn't underrate the skill with which even the backward cells of our own bodies set about making proteins or lipids; they still do better than the scientist knows how to imitate.

But even though there are differences in ability between plants and animals when it comes to synthesis, many general features are possessed in common, and, as we have already come to expect in the context of the energy-providing pathways, a large number of the more important mechanisms are shared by all living organisms. Certain properties characterize all biosynthetic processes Firstly, just as the catabolic systems we described in the last chapter are *providers* of energy, so biosynthetic routes *use* energy, so that much of the ATP produced during oxidation of glucose or of fatty and amino acids is used in the formation of highly reactive phosphorylated intermediates on the pathways for the building of new molecules of carbohydrates, fats, and proteins.

We have already pointed out that these macromolecules are being constantly renewed by the cell, so that even the healthy, fully-grown animal is kept busy the whole time making new molecules. It is difficult if not impossible to assess just how much of the cell's ATP is being used for this purpose. But it is possible to calculate that a typical mammalian cell uses and replaces its entire ATP store once in every 1–2 minutes – or ten million molecules of ATP every second. And we can also be sure that however efficient the utilization of ATP is, 100 per cent efficiency in biosynthesis will not exist any more than 100 per cent of the energy released during biological oxidation could be trapped and exploited by the cell; indeed, we shall see how the cell nearly always has to expend more molecules of ATP in making any particular substance than could be gained by oxidizing it. This, of course, is why we die without food – we need a constant supply of potential energy from outside our own bodies just to keep ourselves ticking over, even before we begin to think about actually doing work on our surroundings.

All biosynthetic pathways, then, are energy-requiring, and in this sense are the reverse of those catabolic pathways we have so far discussed. Many biosynthetic pathways are the reverse of catabolic ones in another sense also. The energy-giving reactions we have described proceed by *oxidation* of their substrates; biosynthesis, on the other hand, demands the *reduction* of its simpler starting molecules as the more complex materials are built up. Just as cellular oxidation was mediated by coenzymes, and in particular by NAD, we shall find that cellular reduction is also coenzyme demanding, and that the key hydrogen donor is, not NAD, but its close relative, the phosphorylated NADP which we have already met in the role of a hydrogen acceptor for a small number of oxidative dehydrogenase reactions.

The reaction $NADPH_2 \rightarrow NADP$ has a $\Delta G°$ of -167 kJ so its use as a hydrogen donor also provides the means by which this exergonic reaction can be linked to an otherwise endergonic reaction to ensure an overall negative free energy change. This then is another molecule behaving in an analogous way to ATP in the cellular energy economy.

Biosynthesis, as should now be apparent, is a more complicated process than oxidation, and is by no means yet completely unravelled. The problems of investigating biosynthesis were initially complicated by the fact that many of the catabolic reactions discussed in the last chapter are reversible. We have already seen how, starting with acetyl-CoA or pyruvic acid, the cell can synthesize glucose by a reversal of the Embden–Meyerhof pathway, so that it was a natural assumption that biosynthetic routes in general occurred by the exact reversal of oxidative ones. Many distinguished investigators, especially in the fields of glycogen and of fat synthesis, were led astray by failing to recognize that though an enzyme might be able to catalyse both a forward and a backward reaction in the test-tube, in the cell it is not likely ever to be given the freedom to do this. Conditions and relative concentrations of the reactants are always likely to favour one direction rather than the other. It would indeed be a very false economy for the cell to attempt both to degrade and synthesize molecules using the same set of enzymes. The confusion that would result would be like that in a busy shop where crowds of people try to climb a staircase that is already full of shoppers coming down. Life is easier with two stairways, marked clearly 'up' and 'down', and it is now clear that in general this is the principle employed within the cell.

This chapter begins to look at some typical examples of 'up staircases', for the synthesis of glycogen, fats, and some of the more complex small molecules the cell uses. We will leave the most complex of the 'up staircases', those for protein and nucleic acid synthesis, to a chapter of their own.

SMALL MOLECULES

The last chapter showed the degree of interaction of the metabolism of small molecules like glucose and the amino acids. Certain cells, though not all, can resynthesize glucose from intermediates of the tricarboxylic acid cycle, fixing carbon dioxide in order to do so. Saturated, but not all unsaturated, fatty acids can be made from acetic acid; the unsaturated fatty acids that cannot be made must be taken in in the diet, and are known as *essential* fatty acids; linoleic acid is an example. Amino acids, too, can be generated from the intermediates of glycolysis and the tricarboxylic acid cycle, although a significant proportion of the amino acids present in protein cannot be synthesized, at least in humans, who lack the necessary enzymes. Hence, such amino acids as tryptophan, phenylalanine, leucine, valine, methionine, lysine, threonine, and isoleucine must also be supplied in the diet and are similarly known as essential amino acids. Serine, glycine, and cysteine can be made from 3-phosphoglyceric acid, provided glutamic acid is present as a donor of NH_2 groups. The glutamic acid itself can be made by a reaction between α-oxoglutaric acid (a tricarboxylic acid cycle intermediate) and ammonia, and participates in the formation of aspartic acid, arginine, proline, and other amino acids. Aspartic acid and glutamic acid also provide the nitrogen for the production of purines and pyrimidines, while the sugar portion of the nucleotides comes by way of ribose-5-phosphate, itself a product of the pentose phosphate pathway. Important and interesting as these metabolic interactions and syntheses are, we will not consider them in any further detail here (they form a large part of the central chapters of many standard textbooks of biochemistry) but pass on straight away to a consideration of the synthesis of the macromolecules.

GLYCOGEN SYNTHESIS

The glucose polymer of animals, glycogen, is made mainly by liver and muscle cells. The starting material for its synthesis may be regarded as any glucose surplus to the immediate energy requirements of the cell, and we can view the glycogen as being made by the binding together of glucose units which have been activated by phosphorylation from the ATP molecule. We previously traced the breakdown of glycogen to provide oxidizable glucose at the hands of the enzyme glycogen phosphorylase (page 181), which, beginning with a chain of glucose units and inorganic phosphate, can split the molecules off from the chain one at a time as glucose-1-phosphate. The glycogen phosphorylase reaction is reversible, and the purified enzyme, if provided with glucose phosphate and a 'primer' of a bit of glycogen chain (it needn't be more than two glucose units long), will perform in the test-tube the reaction of glycogen synthesis – it will begin sticking glucose units on to the primer chain and releasing inorganic phosphate. For a long time it was thought that this really was the way glycogen was made within the cell itself, despite the fact that the equilibrium of the phosphorylase reaction lies far in the direction of glycogen breakdown and not that of synthesis, so that even in the artificial, test-tube situation the glucose phosphate must be present in very high concentrations to drive the synthetic reaction.

However, by the 1960s it became quite clear that, in the living cell, glycogen synthesis followed quite another and more complicated route than the reversal of phosphorylase action, and that in this new pathway the equilibrium of the reaction was shifted decisively away from the sugar phosphate and towards the polymer chain formation. This breakthrough followed the discovery by Luis Leloir in Argentina of a new coenzyme derived from the pyrimidine, uracil. This substance, uridine diphosphate, UDP, can participate in a group transfer reaction with ATP to form the triphosphate, UTP, and it is UTP that is the common intermediate that provides the coupling link between ATP breakdown and the energy-requiring build up of the glycogen chain. UTP, Leloir found, can react with glucose-1-phosphate to give uridine diphosphate glucose, or UDPG for short (1):

$$\text{UTP} + \text{glucose-1-phosphate} \longrightarrow \text{UDPG} + \text{(P)} - \text{(P)} \qquad (1)$$

Glucose, by being converted into glucose-1-phosphate, has already been activated. Now, when it combines with UTP to form UDPG it becomes very reactive indeed. Attached to its UDP coenzyme, the glucose can undergo a whole series of conversions that are otherwise impossible for it; it can, for example, be readily converted into several other hexose isomers of glucose, such as galactose. But from our point of view the most important reaction of UDPG is that catalysed by the enzyme glycogen synthase. This enzyme easily performs the reaction which phosphorylase finds so hard, and transfers the glucose from UDP on to a 'primer' to begin building the glycogen chain:

$$\text{UDPG} + \underset{\text{(primer)}}{\text{gluc-gluc}} \longrightarrow \text{UDP} + \text{gluc-gluc-gluc} \qquad (2)$$

$$\text{UDPG} + \text{gluc-gluc-gluc} \longrightarrow \underset{\text{etcetera}}{\text{UDP}} + \text{gluc-gluc-gluc-gluc} \qquad (3)$$

This reaction can be repeated indefinitely until the chain at last reaches the right length. Meanwhile, the UDP set free is converted into UTP at the expense of ATP:

$$\text{UDP} + \text{ATP} \longrightarrow \text{UTP} + \text{ADP} \qquad (4)$$

Two ATP molecules are therefore consumed for each glucose unit added to the glycogen chain, and not the one which would be the case if the synthesis were in fact catalysed by phosphorylase.

This does not quite complete the story of glycogen synthesis, because, as we know (page 53), the glycogen molecule is not straight-chained but branched, and both the UDPG system and the phosphorylase one are only capable of polymerizing glucose into straight chains linked by 1—4 bonds between the glucose units, The branching points are, it will be recalled, made by joining one length of 1—4 linked chain to another by a 1—6 linkage. The links are provided by the action of the *1, 4 α-glucan branching enzyme* which works by a highly specific mechanism so as to form the branch points in exactly the same way each time. The enzyme removes a fragment of about seven glucose residues from the non-reducing end of a linear chain (that is, the end that has a free hydroxyl at C4 rather than C1) and attaches the fragment to the hydroxyl group on C6 of any glucose molecule in the third place in from the non-reducing end of another chain. The repeated concerted action of glycogen synthase and the branching enzyme results in the formation of the very characteristic highly branched glycogen molecules each of which contains about 200 to 300 glucose units in all, to be deposited as short-term food stores in the cells of the liver and muscle.

Glycogen synthesis in many ways provides a general model for polysaccharide biosynthesis, because in every one of those pathways so far studied nucleoside diphosphates like UDP seem to act as the glycosyl carrier, donating glycosyl molecules to an incomplete polysaccharide chain.

FATTY ACID SYNTHESIS

The problem of fatty acid synthesis, like that of glycogen, is an example of biochemists being led astray, by the relative ease with which the oxidative pathway for fatty acid breakdown was unravelled, into believing that the synthetic route was simply its reverse. The oxidation of fatty acids by the enzyme system of pages 182–7 proceeds by way of alternate hydration and dehydrogenation of the coenzyme A derivatives of the acids, in a sequence of repeating reactions which chip away 2-carbon fragments from the end of the carbon chain and release them as acetyl-CoA. All the reactions in the sequence are reversible, at least in the test-tube, and for a long time it was believed that this reversal was the general mechanism responsible for the manufacture of the fatty acids in the living cell.

However, eventually certain important differences began to emerge. For instance, the oxidative pathway uses the coenzyme NAD, whilst when the purified enzymes of the synthetic system were studied, it was found that they worked better with $NADPH_2$ as coenzyme than with $NADH_2$. So it was proposed that the synthetic system was the reversal of the oxidative system with one difference – that NADP replaces NAD at the critical reduction step:

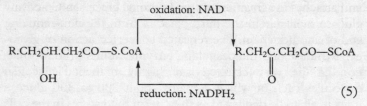

$$\text{(5)}$$

But this proved to be only the first of a set of differences from the degradative pathway that are required for effective synthesis of fats. It was, for instance, observed that cell preparations which had been made in solutions which contained the bicarbonate ion HCO_3^-, a ready source of dissolved carbon dioxide, made fatty acids much better and faster than those that were made in non-bicarbonate solutions. This immediately suggested that in some way fatty acid synthesis required carbon dioxide, yet, when experiments were made using radioactive CO_2, none of the label appeared in the fatty acids. The effect of the CO_2 appeared to be a catalytic one. Most important, though, was a theoretical problem similar to that of glycogen synthesis. The last of the reactions of fatty acid oxidation is that of thiolysis by β-ketothiolase (6):

$$\text{R.CO.CH}_2\text{.CO.SCoA} + \text{HSCoA} \rightleftharpoons$$
$$\text{R.CH}_2\text{.CO.SCoA} + \text{CH}_3\text{.CO.SCoA}$$

$$\text{(6)}$$

(see also equation (31), page 185).

Like the phosphorylase reaction in glycogen metabolism,

although thiolysis is theoretically reversible, the equilibrium lies far over to the right; in order for the cell to use it to build fatty acids it has to collect a formidably high concentration of acetyl-CoA and R. $CH_2.CO.SCoA$. Again, a reaction possible in the test-tube is by no means so easy for the cell in practice. It is clear that all these facts, taken together, point to the existence of an alternative route of fatty acid synthesis.

The clue to the new pathway was provided in 1958 by Salih Wakil, in the USA. Wakil sought a system whereby the cell could bypass the unfavourable equilibrium of reaction (6), and found that the trick was done, just as by UDPG in glycogen synthesis, by an activation of the reactants. Once again, activation is provided at the expense of ATP and a new coenzyme. What the cell does is to convert acetyl-CoA into another, more reactive substance, malonyl-CoA. Malonic acid is produced by the addition of a molecule of carbon dioxide to acetic acid, (7):

$$CH_3.COOH + CO_2 \longrightarrow CH_2 \begin{array}{l} \diagup COOH \\ \diagdown COOH \end{array} \qquad (7)$$

and it is obvious from its formula that the molecule will be more unstable, and thus more reactive, than that of acetic acid.

The enzyme responsible for carbon dioxide fixation is called acetyl-CoA carboxylase and is dependent upon the presence of the metal ion manganese and also the vitamin biotin. In fact it is the biotin bound to an enzyme complex (Biotin-E) that first reacts with the CO_2 in an energy-requiring reaction that is driven by the simultaneous breakdown of ATP. The carboxybiotin E which is formed can now be thought of as an activated carrier molecule, very similar to UDPG. It can easily transfer its CO_2 to acetyl-CoA to give malonyl-CoA in the following reactions:

$$\text{Biotin-E} + CO_2 + \text{ATP} \longrightarrow \text{carboxybiotin-E} + \text{ADP} + \textcircled{P}_i \qquad (8)$$

$$CH_3CO.SCoA + \text{carboxybiotin-E} \longrightarrow CH_2 \begin{array}{l} \diagup COOH \\ \diagdown CO-S.ACP \end{array} + \text{biotin-E} \qquad (9)$$

The next step in the pathway is the linking of malonyl-CoA and acetyl-CoA to different molecules of another sulphydryl-containing protein, acyl carrier protein (ACP). It is on their ACP derivatives that malonyl-CoA and acetyl-CoA combine to give the β-keto acid acetoacetic acid, still linked to ACP, but releasing CoA and CO_2.

$$CH_3-\overset{\overset{O}{\|}}{C}-S.ACP + CH_2 + \underset{\underset{CO-S.ACP}{}}{\overset{\overset{COOH}{}}{CH_2}} \longrightarrow$$

$$CH_3-\overset{\overset{O}{\|}}{C}-CH_2-\overset{\overset{O}{\|}}{C}-S.ACP + ACP-SH + CO_2 \qquad (10)$$

Thus carbon dioxide is alternately fixed by reaction (8) and released once more by reaction (10); it is in fact acting catalytically, as expected from the experiments made in the presence of bicarbonate that we have already mentioned.

The net result of reactions (8)–(10) is the linking of two molecules of acetyl-CoA to give one of acetoacetyl-CoA. As the equilibrium of the reactions is favourable to synthesis, we have overcome the stumbling block provided by the β-keto-thiolase reaction of equation (6). Once again, reversal of equilibrium has been bought at the price of a molecule of ATP. The acetoacetyl-CoA is now free to react with another molecule of malonyl-CoA, thus building up large fatty acid chains.

Thus the overall reaction for the production of a molecule of the fatty acid palmitic acid from acetic acid is as follows:

$$\begin{array}{c} AcetylCoA + 7 \text{ malonylCoA} + 14 \text{ H} \longrightarrow \\ \text{palmitic acid} + 7CO_2 + 14 \text{ NADP} + 6H_2O \end{array} \qquad (11)$$

In addition, 23 ATP molecules will have been utilized, 16 to convert 8 pyruvic to 8 acetyl-CoA and 7 to carboxylate 7 acetyl-CoA to 7 malonyl-CoA molecules.

Thus although the cell probably makes some of its fatty acid by simple reversal of oxidation, the larger part of it is produced by a system of enzymes which, although performing formally similar reactions, in fact differ in certain essential respects. In particular, they use NADP instead of NAD as coenzyme, and bypass the difficult thiolysis reaction by means of new intermediate, malonyl-CoA, at the expense of an extra molecule of ATP. This mechanism appears to be universal, in bacteria, plants and animal cells. In the animal cell, the entire enzyme system is combined as a multienzyme complex, *fatty acid synthase*, located in the cytoplasm and consisting of two identical subunits each with a molecular weight of 240 000, and containing all the enzyme activities required to take in the small starter molecules at one site and not release them until a complete fatty acid like palmitic has been synthesized. Once palmitic acid has been made, it can be lengthened to make stearic acid, and, in the presence of oxygen, $NADPH_2$ and an enzyme system present in the endoplasmic reticulum, can be *desaturated* to insert a double bond, thus converting it into oleic acid (page 78). Animal cells cannot produce further desaturations, and more unsaturated fatty acids, like linoleic, which are essential, must be obtained in the diet, ultimately from plants, which can synthesize them.

TURNING FATTY ACIDS INTO FATS

Fats are converted into fatty acids by hydrolysis under the influence of esterases (page 114). But, once again, though the esterase system is readily reversible, the route adopted in biosynthesis is more complex, and demands an energy supply in the form of ATP and coenzyme A. At least two ways of making triglyceride fats from the fatty acids have been found. In one, the glycerol is first activated in a now familiar manner, by phosphorylation to give α-glycerophosphate:

$$
\begin{array}{llll}
CH_2OH & & CH_2OH & \\
| & & | & \\
CHOH & + \ ATP \ \rightarrow & CHOH & + \ \ ADP \qquad (12)\\
| & & | & \\
CH_2OH & & CH_2O\!-\!\!\!\textcircled{P} &
\end{array}
$$

The activated glycerol can now react with first one and then a second molecule of palmityl-, stearyl-, or oleyl-CoA:

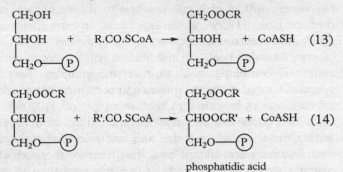

phosphatidic acid

The product is phosphatidic acid, itself an important substance (see page 80) which we will meet again in the context of the cell signalling and messenger systems described in Chapter 11, as well as being the starting point for the phosphatides and related lipids. The conversion of phosphatidic acid into triglyceride demands first the removal of the phosphate through hydrolysis by the enzyme phosphatidic acid phosphatase (an irreversible reaction). The product is diglyceride, which can react with a final fatty acyl-CoA to yield triglyceride:

$$\begin{array}{c} CH_2OOCR \\ | \\ CHOOCR' + H_2O \longrightarrow \\ | \\ CH_2O\!\!-\!\!\textcircled{P} \end{array} \qquad \begin{array}{c} CH_2OOCR \\ | \\ CHOOCR' + \textcircled{P}i \\ | \\ CH_2OH \end{array} \qquad (15)$$

phosphatidic acid   diglyceride

$$\begin{array}{c} CH_2OOCR \\ | \\ CHOOCR' + R''.CO.SCoA \longrightarrow \\ | \\ CH_2OH \end{array} \qquad \begin{array}{c} CH_2OOCR \\ | \\ CHOOCR' + CoASH \\ | \\ CH_2OOCR'' \end{array} \qquad (16)$$

diglyceride   triglyceride

Four molecules of ATP are consumed in making fats by this route; one for making glycerophosphate and the other three for the CoA derivatives of the fatty acids. This reaction path is the one followed in adipose tissue where fat deposits are laid down for the body as a whole. But another slightly more economical route occurs within such cells as those of the intestinal wall, kidney, and liver, where the monoglyceride monolein can react directly first with one and then with another fatty acyl-CoA molecule to give a triglyceride without the need for the prior phosphorylation of glycerol. The enzyme responsible for these reactions is monoglyceride transacetylase, which thus makes triglyceride using only three ATPs.

But even with this marginal saving in the efficiency of esterification, the whole process of fat synthesis by the cell is an object lesson demonstrating how much harder it is to build than to destroy, a lesson we will find repeated again as we pass to the synthesis of even more complex molecules than fats and polysaccharides.

CHOLESTEROL SYNTHESIS

The other class of lipids described in Chapter 3 is the steroids, the starting point for the production of which is the 27-carbon molecule cholesterol. Cholesterol is synthesized in all animal cells, and some also enters the body by way of the diet. Cholesterol is an essential component of membranes, especially nerve cell membranes, but an excess either through dietary intake or synthesis can lead to high levels in the circulating blood. Cholesterol can then become laid down in the membranes of arteries, leading, potentially, to their blockage and hence to arterial diseases such as coronary heart disease. The molecule has a complex ring structure which can be drawn as shown below.

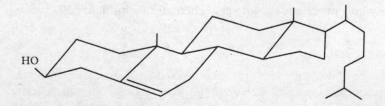

Although the structure of cholesterol was identified in the 1930s, it took until the 1970s to work out its synthesis. It is derived from acetoacetyl-CoA and acetate, and hence by the degradation of saturated fatty acids. This is why one way of reducing cholesterol in the bloodstream is to reduce the dietary intake of saturated fatty acids (as well as cholesterol itself). This does not always resolve the problem of high levels of blood cholesterol, as some people have a genetic condition in which excess cholesterol is synthesized, unless one or more of the enzymes which regulate its production can be blocked by drugs. These mechanisms were uncovered by M. S. Brown and J. L. Goldstein in the US in the 1970s and 1980s, who were recognized by the award of the 1985 Nobel Prize. Statins, drugs which inhibit a key CoA-dependent synthetic step in cholesterol production, are now much favoured for the treatment of high blood cholesterol levels.

The synthetic problem is how to go from what are essentially straight-chain molecules like the fatty acids to the ring molecules of which cholesterol is the prototype. This requires more than just simple addition and subtraction of carbon units but means that the long, straight-chain molecule itself must be curled round and joined up to itself like a snake swallowing its tail. No fewer than twenty-four reaction steps are involved in three main stages. In the first, acetyl-CoA is combined with acetoacetyl-CoA to give a new, 6 carbon intermediate, *mevalonic acid*. In the second set of reactions, mevalonic acid is further activated by phosphorylations and three molecules of the acid are linked together, via a complex set of intermediates, to give 1 molecule of *squalene*. Squalene is then converted through a further intermediate, *lanosterol*, to cholesterol itself. The whole process is shown very schematically in Figure 27.

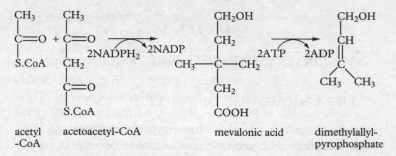

acetyl -CoA   acetoacetyl-CoA   mevalonic acid   dimethylallyl-pyrophosphate

Drawing the last molecule in skeleton form, its coalescence to eventually form cholesterol can be shown as:

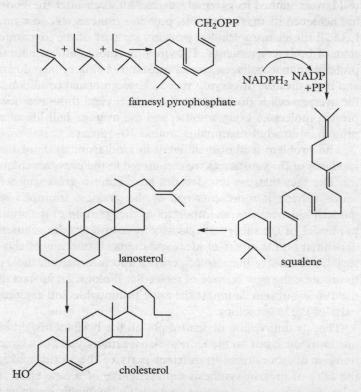

Figure 27. Formation of cholesterol

CHAPTER 10

Synthesis of Proteins and Nucleic Acids

THE UNIQUENESS OF PROTEIN SYNTHESIS

Protein synthesis is one of the most important of the cell's synthetic activities. The growing animal makes protein at an extremely rapid rate; up to seventy per cent of the liver of a growing rat can be removed, and within twelve days the organ will have regained its original weight. But even after the body has achieved its maximum size, protein synthesis still goes on. Like all the giant molecules, proteins are part of the 'dynamic state of body constituents'. Enzymes, haemoglobin, structural proteins such as collagen, are all constantly being broken down and resynthesized at varying speeds. Under normal conditions, the average cell is probably synthesizing several thousand new protein molecules every minute, and the average half-life of a protein in an adult mammal is around 10–15 days.

The problem is almost different in kind from those of the synthesis of the substances we considered in the previous chapter. The fact that the last decades have gone a great way towards solving it represents one of the greatest triumphs of modern biochemistry, as important as the splitting of the atom in physics or the theory of relativity in cosmology. Its solution depended on the inflow of ideas and methods not only of classical biochemistry but also of genetics and information theory to produce the new science of molecular biology, an upstart in the 1960s but now amongst the most fashionable and exciting parts of the new biology.

That its unravelling depended not on the brilliant insight of one individual, but on the hard work over many years of a large number of laboratories in different parts of the world, makes the saga of protein synthesis representative of a new class of scientific discovery – one that depends on the highly organized state of modern science, so that workers in different disciplines and different continents are kept constantly in touch with one another's experiments and thoughts not only by special journals set up to cope with the flow of results, but by conferences, letters, Fax machines, emails – even the daily newspapers.

Why is protein synthesis so different from that of polysaccharides or lipids? The actual chemistry of the formation of peptide bonds between amino acids is a type of reaction that we are already quite familiar with. Two amino acids can link together with the removal of water to form a dipeptide:

$$\underset{\substack{| \\ NH_2.CH.C}}{\overset{R}{|}} \underset{\substack{\\ O}}{\overset{OH\ H}{}} + \underset{\substack{\\ H}}{N.CH.COOH} \longrightarrow \underset{\substack{| \\ NH_2.CH.C}}{\overset{R}{|}} - \underset{\substack{| \\ O\ H}}{\overset{R'}{N.CH.COOH}} + H_2O$$

This reaction is the reverse of those carried out by enzymes such as trypsin and pepsin that hydrolyse proteins during digestion but it cannot be carried out by them because, like other synthetic reactions, it is energy-requiring, and we may anticipate that it will be necessary to provide this energy in the form of ATP. It is possible to test this hypothesis by studying the synthesis of some of the simple peptides that the cell regularly makes.

One of these is the tripeptide glutathione, γ-glutamylcysteinyglycine, a substance found in most cells although its actual function is not clear. Glutathione is made by the cell in two stages: in the first glutamine and cysteine condense to give γ-glutamylcysteine, and in the second the dipeptide reacts with glycine to give γ-glutamylcysteinylglycine. As expected, two molecules of ATP are split during this synthesis.

But the problem of protein synthesis is more complex than that of providing an enzymic mechanism for peptide bond formation. Glycogen synthesis was solved when we demonstrated an enzyme system which could add glucose on to the growing polysaccharide chain, and a branching enzyme to fork the chain at fixed intervals. But glycogen is made of only one sort of sugar – glucose. Even more complex polysaccharides contain only three or four different sorts of sugar at most, bound together in a repeating pattern in which the exact sequence

211

may not be too vital. By contrast the protein chain is made of twenty different sorts of amino acid joined head to tail in a unique and non-repetitive sequence. Each amino acid has its predetermined place in the sequence, and altering it in some way, by the replacing of one amino acid by another, or by the inverting of the order of two or more along the chain length, may mean that the protein will no longer fulfil its physiological function within the cell (see, for example, page 39). The protein synthesis problem is one of specificity.

The problem, as it might have appeared about 1950, say, can thus be summarized as that of how to assemble a series of up to 100 amino acids head to tail along a peptide chain so that the order of the amino acids along the chain could be accurately predetermined and reproduced.

It is possible to visualize two ways of building up such a chain. Either we could start with one amino acid, add a second to it, then a third, a fourth, and so on, until the chain is complete; or, alternatively, we could collect together all the amino acids needed for the chain, and, when they were all assembled, link them almost simultaneously one to another (like doing up a zip-fastener). The second alternative seems an unlikely one, and would not be considered at all were it not that the first leads into a morass of difficulties. Consider a hypothetical example. To join glycine to alanine, we can envisage an enzyme which performs the reaction:
(like the glutathione synthesis described earlier). Now imagine

$$gly + ala \longrightarrow gly\text{-}ala$$

a second enzyme to add a third amino acid – say tyrosine:
To add a fourth amino acid we need a third enzyme, and to add

$$gly\text{-}ala + tyr \longrightarrow gly\text{-}ala\text{-}tyr$$

a fifth, a fourth enzyme is required. In general, to build a chain of n amino acids, $(n-1)$ enzymes would be required. But as the protein chain grows in length, so the amount of information that the nth enzyme needs in order to add the $(n+1)$th amino acid also increases. In deciding whether to add serine as the twenty-fifth amino acid, the enzyme must have at its active centre a means of checking all the preceding twenty-four acids to ensure that they are present in the right order. Now enzymes may be good at the job, but this must obviously impose an enormous strain upon them. By the time the 99th enzyme adds the 100th amino acid, error is bound to have crept in. And errors, as we know, are likely to prove fatal. But there is a still greater absurdity, for we have postulated a process whereby, to make a single protein chain of 100 amino acids, 99 enzymes are required. As there are likely to be some 10 000 different proteins present in the cell, to make them we need a total of 990 000 specific enzymes! And to make each of these 990 000 enzymes, each of which is a protein, we need another 99 enzymes ... Clearly the argument is a nonsense.

Experimental evidence backs up theory in ruling out this stepwise enzymic synthesis of protein. If proteins were made by a steady build-up of peptide subunits, study of a cell that was engaged in active protein synthesis should reveal the presence of many of such peptide fragments. Yet few have ever been found, and those that are found are always specific molecules with particular biochemical roles themselves – like glutathione, for instance – and not merely precursors of completed protein chains. Similarly, if a tissue is fed all but one of the amino acids essential for synthesis, it would be expected that a large number of peptide chainlets would be formed, representing growing protein chains which had been broken off where the missing amino acid should have fitted in. But in fact under such conditions what happens is that synthesis stops completely, and no fragments are ever found. Thus the hypothesis that proteins are formed from amino acids by stepwise enzymic synthesis must be abandoned.

TEMPLATES AND GENES

Yet the alternative hypothesis, as it stands, seems equally unsatisfactory. It is difficult to picture a system in which 300 or so free amino acids come spontaneously together in the correct order to link up to form a chain. The odds against it would be immense. Some method must exist whereby the amino acids are assembled at particular sites, held in waiting until all are present and correct, shuffled into an approved order, and then linked. This implies the existence of 'sites' at which the amino acids can be 'parked' pending the completion of the chain. Each site must be able to distinguish between the different amino acids so that only an acceptable one can be fitted in, and the sites themselves must be linked to each other in the same order as the amino acids of the final protein chain. In other words, protein synthesis demands a pre-existing set of 'amino acid recognition sites' forming part of a pre-existing macromolecule.

Such an answer to the riddle satisfied biochemists, geneticists and information scientists alike. A 'mould' or 'template' for protein synthesis was predicted by the biochemical requirements, whilst, as will become clear shortly, if it could be shown that the template was under genetic control, the clue to how information on protein structure is transferred from cell to cell would be found. What would be the requirements of a template? It must be a macromolecule related in length to the protein it is synthesizing. Along its length must be a series of sites corresponding to the individual amino acids of the protein, and each such site must be capable of distinguishing between different amino acids and accepting only one. There are only three types of macromolecular chain: carbohydrates, proteins, and nucleic acids. Carbohydrates can be ruled out because chains, say, of glucose in glycogen cannot possess the high degree of specificity demanded of a structure designed to distinguish between amino acids. Nor can it be simply that protein chains are made on other proteins as templates. Although, evidently, a protein fulfils the criteria of specificity we have laid down, it would mean that proteins were a self-reproducing set of molecules. This gives the same sort of logical absurdity found when we considered the possibility of a stepwise enzymic synthesis.

There remain the nucleic acids as possible templates. DNA and RNA are macromolecular chains often many times greater in length than proteins. They are composed of a set of four different bases arranged in a complex but ordered pattern along their chain length. Such bases could easily be organized into a sequence of 'sites' that could discriminate between amino acids, and could thus fulfil the criteria we have laid down for the template. Thus the proposition that the nucleic acids form a template on which proteins are synthesized makes biochemical sense.

It also makes genetic sense, although it wasn't until the early 1950s that this became quite clear. Geneticists are concerned with how cells and organisms reproduce themselves. Classical genetics started in the late nineteenth century with Gregor Mendel, who proposed laws to account for the regularities he found when he studied how plants such as peas crossbred to produce varieties of different colours and shapes. Modern genetics, a science which grew up practically contemporaneously with biochemistry and molecular biology, is concerned with two main questions. First, what is the origin of the extraordinary *similarities* between individuals of the same species, so that, starting with the fusion of a mere two cells, ovum and sperm, cells multiply and organize themselves into individuals that resemble each other so closely in biochemistry and form? The proteins of the infant are largely identical with those of his or her parents. Therefore sperm and ovum together must have packed within them detailed information relevant to the precise manufacture of all the several thousand different proteins in the billions of different cells of the infant's body, and to the construction of that body in recognizable form, with the right number of arms, legs, internal organs and so forth. But an equally important question for genetics is that of the origins of differences between individuals of the same species and indeed of the same parents. Why is one child brown-eyed and another blue-eyed, for instance; why does one have normal and another sickle-cell haemoglobin? To both types of question, understanding the mechanism of reproduction, of cells and of their proteins, is central.

In some ways this question can be considered one of

information storage and transmission. The mechanism for ensuring the exact replication of a protein chain by a new cell is that of transferring the information about the protein structure from the parent to the daughter cell, and subsequently of translating that stored information back into the synthesis of a protein chain again. Geneticists had known for many years that when a cell divides, or two cells fuse in sexual reproduction, the information that the new cell needs for controlling its growth and development is contained in a small number (in humans, 23 pairs) of thin strands of material which appear under the microscope like a handful of twisted streamers after a child's party. Those strands are called *chromosomes*. When a cell reproduces by division, each chromosome splits lengthwise into two halves, so that each daughter cell contains the same number of chromosomes as its parent. In sexual reproduction, each parent cell (sperm or ovum) provides half the chromosomes of the offspring. Thus, whether produced sexually or by cell division, every new cell ends up with the same number of chromosomes as the parent cell.

Before the advent of molecular biology, geneticists conceived each chromosome as consisting of many thousands of units strung along its length like beads on a necklace. Each unit, or bead, contains information relevant to the expression of a particular characteristic property which, as the cell develops, will unfold in interaction with its environment. Characteristics such as colour of hair or eyes, or for normal or abnormal haemoglobin, are all examples of the types of property possessed by an organism (its *phenotype*, as geneticists term it), for which the relevant genetic information may be found within the chromosome. The beads, or genetic units, are called *genes* by the geneticists, who deduced their existence theoretically long before the electron microscope enabled the chromosome to be examined visually in detail and the genetic units along its length mapped by a set of genetic techniques such as the use of controlled mutation in bacteria – and more recently the recombinant technology discussed in Chapter 15. Although this view of the role of genes and chromosomes is now known to be oversimplified – indeed in some ways misleading – it served to colour the thinking of those setting out in the middle of the twentieth century to solve the protein synthesis riddle. The total complement of genes in an organism is known as its *genome*.

In the biochemical picture of the cell that we have built up, the day-to-day running of the cell is the job, not of genes, but of specific proteins; in particular, enzymes. Eye or hair pigments, for instance, are actually made by a series of enzymes, even though the information about how to make them is contained in the genes. Thus the role of the genes came to be seen by geneticists as organizers and controllers of protein synthesis. Studies on mutations in moulds by George Beadle and Edward Tatum in Stanford, California, led them by the 1940s to formulate their famous hypothesis: 'one gene – one enzyme'; that is, that each gene was responsible for controlling the production of a particular enzyme, or other protein. Although this is now known to be a great oversimplification, for reasons that will become apparent shortly, this hypothesis helped point the way towards the relationship between genes and proteins which the next fifty years of research would unravel.

But the crucial question was then to discover what genes were made of. Chromosomes consist of a tight complex of DNA and a particular class of very basic proteins, the histones (page 69). The histones are arranged in units of eight molecules (octomers), together with a length of DNA about 200 nucleotides in length, so that under the electron microscope the chromosomes have a beaded appearance. Each of these beads (composed of a histone octomer plus its associated DNA) is known as a *nucleosome* and combinations of these nucleosomes together form genes.

For a long time it was believed that it was the proteins that were the vital information-bearing component, and it took a series of experiments through the 1940s and early 1950s to break the power of this protein-oriented thinking amongst biochemists and geneticists. Although evidence that the genetic material is nucleic acid was first proposed by a medical doctor,

Frederick Griffith, in London in 1928, it was Avery and his colleagues MacLeod and McCartney at the Rockefeller Institute in New York in 1944 who purified DNA free of protein and showed chemically that in both bacteria and eukaryotes, the genetic material is indeed made of DNA. This discovery offered a unifying view for the basis of heredity, although there are exceptions: for instance, in some viruses the genes are built of RNA instead. Alfred Hershey and Martha Chase, in 1951, labelled the nucleic acid of viruses with radioactive phosphorus, and the protein with radioactive sulphur, and then allowed the viruses to infect bacterial cells. The phosphorus, but not the sulphur, entered the infected cells as the viruses multiplied inside them; thus the virus must work by injecting its nucleic acids into the cell it was attacking, leaving the protein outside. At that time Avery made a comment with wider-ranging implications than he could have anticipated – that perhaps genetic material could be transferred between different species and yet remain functional.

Nearly all of these key experiments were made using bacteria and the viruses that infect them. A virus is to all intents and purposes a packet of nucleic acid wrapped in a protein jacket. It exists by attacking cells – animal, plant, or bacterial – taking over the host's protein and nucleic acid synthesizing systems and using them to make more viral protein and nucleic acid. With these it replicates its own structure several times, until it has used up and exhausted all the utilizable substrates present in the host cell. Finally, the virus bursts the host cell, releasing several new virus particles to hunt for fresh prey (see Plate 8). Viruses that attack bacteria are called *phages* (or bacteriophages). One of the most exciting features of this work was the perhaps surprising discovery that the mechanisms which were uncovered in bacteria and their viruses were of universal significance, and with minor differences could be shown to occur in plants and animals as well (for instance, as mentioned in Chapter 4, bacteria do not have nuclei, and therefore bacterial DNA, in its chromosomes, is dispersed through the cell rather than being confined to a single organelle). Bacteria were particularly useful for experiments of this type though because of the ease with which specific mutations can be produced in them by irradiation with X-rays or by chemicals. Each single mutation, it is now known, corresponds to the addition, deletion or substitution of a specific base along the length of the DNA chain, and it is these changes in the DNA sequence which ultimately result in the production of proteins with faulty amino acid sequences or other deviations from normality.

The key transition from thinking about protein to thinking about DNA, however, came not so much from experiment (it could always be claimed that DNA preparations which seemed to carry information were contaminated by small amounts of protein that were really responsible) but from theory: the discovery of the DNA double helix by Watson and Crick in 1953. The base-pairing of the DNA double helix means that, unlike protein, one strand of the DNA does provide a copying mould or template for the other and, as Watson and Crick noted at the end of their paper describing their proposed structure, therefore immediately suggests a mechanism for replication during both cell division and reproduction. To see why, it is necessary to postpone discussion of protein synthesis for a while, and consider the mechanism and significance of DNA synthesis.

DNA REPLICATION

Although earlier we argued that, as far as proteins were concerned, one protein could not act as a mould or template for another, this argument does not apply to DNA. DNA synthesis, it turns out, is *not* the synthesis of a molecule from its constituents like that of glycogen or the fats, but the direct 'copying-off' of one DNA molecule from another, a sort of photocopying process known as *replication*. It is the unique structure of DNA which makes this photocopying possible. Although in popular accounts DNA is often described as a molecule that is 'self-replicating' as if it can simply make copies of itself unaided, as we shall see, this is a major misconception. Making copies of DNA, and then later reading off those copies to make RNA and protein, is one of the most subtle, highly orchestrated and controlled activities that the cell undertakes, and certainly can't be done by DNA alone. Indeed, in the test tube DNA molecules are quite inert. What brings them to functional life is the enzymic and structural environment in which they are embedded.

DNA, it will be recalled (and it is worth looking back at the figure on page 75) is a double-stranded or duplex molecule composed of two chains of nucleotides held together by hydrogen bonding between the purine and pyrimidine base pairs. During replication, which occurs when the cells divide, the hydrogen bonds are broken and the two chains separate and unwind. Replication involves a major disruption of the structure of DNA. However, it is only transient and is reversed as the daughter strands are formed. Moreover, only a small part of the DNA loses its duplex structure at any one time. Each chain acts as a template on which a new DNA chain can be synthesized from the constituent nucleotides, which become attached in the requisite order because there are no alternative options open for them except for A to pair with T and G with C. Thus replication is *semi-conservative*, each new DNA molecule, and hence each new cell, containing one of the parent chains plus one new chain, a relationship first proved, using isotopic tracers, by Meselson and Stahl in 1958 (Figure 28).

The enzymes responsible for synthesis of the new chain are *DNA polymerases*, which use as their substrates the deoxyribonucleotide triphosphates. The process is initiated by formation of a junction between the two regions where the double helical structure is disrupted. This structure is called a *replication fork* (Figure 29). Replication involves unwinding of the parental strands, so there is a continuous movement of the replication fork along the parental DNA. Because DNA molecules are so long that simply unravelling the helix from one end would set up an intolerable strain on the rest of the molecule long before the task was completed, synthesis does not simply begin at one end and go on until it reaches the other. Instead it takes place simultaneously in different parts of the molecule

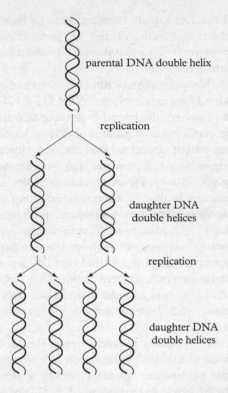

parental DNA double helix

replication

daughter DNA
double helices

replication

daughter DNA
double helices

Figure 28. Replication of DNA

and, to make it easier for the polymerases to recognize where
they should begin replication, the DNA helix has a number of
sites at which replication originates. However, the double helix
of DNA must be first opened up with the aid of *helicase*, an enzyme that uses the energy of ATP to open up the double helix,
and *initiator proteins* that bind to the DNA. The replication fork
is kept open by proteins which bind to the individual DNA
chains, thus preventing formation of hydrogen bonds between
the two strands. DNA polymerase is firmly attached to the single-stranded DNA by a protein that forms a ring around the
DNA chain and is therefore called a *sliding clamp*. Replication
forks formed at multiple replication origins open up the two

strands of the DNA in both directions. In bacteria the forks move at about 1000 nucleotide pairs per second; humans do this much more slowly, moving at the rate of only 100 nucleotide pairs per second.

So, the DNA polymerases not only need to possess binding sites for four different substrates (ATP, GTP, CTP, and TTP), but also have to have the capacity to recognize and bind to the sites where replication originates and they have to locate the base on the parent strand so that they can insert the correct matching base into the growing daughter strand. As well as everything else it does, DNA polymerase has one further important function – that is an error-correcting activity called *proofreading* (it is interesting how often in describing processes involved with DNA and protein synthesis molecular biologists find themselves adopting terms from the publishing lexicon!). It is therefore not surprising to find that DNA polymerases are large allosteric enzymes. One of them, DNA polymerase1, has been shown to contain at least two distinct enzyme activities, polymerization and nuclease activity, within a single polypeptide chain.

Yet another synthetic problem lies in the fact that the two DNA strands in the double helix run in opposite directions – in one chain the nucleotides are linked by a phosphate group through the hydroxyl at carbon atom 5 (C5) of a ribose sugar molecule to the hydroxyl on C3 of the next sugar in line; in the other chain the linkage is from C3 $\rightarrow$ C5. DNA polymerase can only synthesize new chains in the 5 $\rightarrow$ 3 direction (page 71), adding new nucleotides to the 3' end of the chain. What happens is that the enzyme copies one side only at first in the 5 $\rightarrow$ 3 direction leaving one strand unpaired until there is enough room to enable another polymerase molecule to initiate replication of the unpaired chain in the correct chemical direction. This is made discontinuously by DNA polymerase working backwards from the replication fork.

A further problem is that even at the initiation points of replication the polymerase cannot begin *de novo* synthesis unless there is a *primer* – a very short length of single-stranded ribonucleic acid (itself synthesized by a *primase* enzyme) to which the incoming nucleotide can be attached. This RNA primer is about 10 nucleotides long and attaches to the template chain, thus providing DNA polymerase with a starting point for polymerization. The DNA polymerase will continue to synthesize new DNA until it collides with the next RNA primer. A continuous new 3'-to-5' DNA chain is thus made of many separate fragments of DNA, and three further enzymes are needed to complete the job. A *nuclease* will quickly remove the RNA primer, a DNA polymerase called *repair polymerase* will replace the RNA with the DNA and finally, the *DNA ligase* will join the 5'-phosphate end of one daughter DNA fragments to the 3'-hydroxyl end of the next fragment.

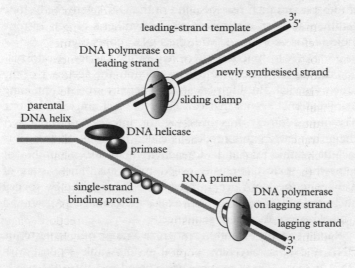

Figure 29. DNA replication fork

It is not surprising that DNA synthesis requires a considerable input of energy. We have already said that the substrates for the polymerase are the nucleotide triphosphates, but it is the monophosphates that are inserted into the chain. So once again the splitting off of the two terminal phosphate groups from the nucleotides provides the driving force for a biosynthetic reaction. This whole process provides an elegant method for the copying of the genetic material, a model which economically and precisely accounts for the molecular events underlying genetic transmission of information across generations. The process of replication by template-directed enzymes avoids the trap of an infinite regress of enzymes to fabricate a macromolecule with a defined structure with accuracy and efficiency, and provides the basis on which we can return to the question of protein synthesis.

As already mentioned, in eukaryotes, as opposed to prokaryotic bacteria, DNA is present in the form of chromatin, a complex of DNA and histones. As a consequence, eukaryotic DNA replication requires two additional steps: dissocation from the chromatin and then reassociation of the DNA with the histone octomers to form nucleosomes. The synthesis of new histone molecules is coordinated with the DNA replication.

For DNA to function effectively as the carrier of information during cell division and reproduction, the accuracy of the process is vital. But life persists in an environment containing many external agents that can enter the cell and convert one base into another, thus breaking the integrity of the DNA chain. In genetics this process is called *mutation*. Such agents include cosmic rays and radioactivity, and also a number of somewhat toxic chemicals, called mutagens. These sites of damage to the DNA are, in most cases, recognized by special nucleases, repair enzymes that excise the damaged region and synthesize a replacement sequence.

Repair systems can often recognize a range of injuries to the DNA and cells may have several systems able to deal with DNA damage. *Direct repair* is widespread in nature, especially in plants, and involves reversal or simple removal of the damage. *Excision repair* probably handles most of the damage that occurs in a cell and is initiated by a recognition enzyme activated either by a change in the spatial organization of the DNA or by an actual damaged base. *Mismatch repair* systems scrutinize DNA for apposed bases that do not pair properly. This system is very active during replication. *Tolerance systems* provide a means for a cell to cope with the difficulties that arise when normal replication is compromised at a damaged site. Repair systems have been studied in great detail in bacteria, but their biochemical characterization is less extensive in eukaryotic cells.

Genes and genomes are often regarded as very stable, changing only slowly during the process of evolution. And indeed, without genetic recombination, variability would occur only as a consequence of mutation that escaped the eye of repair systems. However, recombinations can occur between precisely corresponding sequences of DNA, resulting in direct exchange of material between the DNA duplexes. Recombinations are initiated by a double-strand break in the DNA and require very precise topological manipulation of DNA, achieved by *topoisomerases*, enzymes that may relax or introduce supercoils in the DNA. Another cause of variability is provided by transposons, transposable sections of DNA that are mobile and able to relocate along the DNA chain. It is the existence of these elements, first observed as long ago as the 1930s in maize chromosomes by Barbara McClintock, but regarded as incredible by other geneticists for getting on for forty years, which helped change the conventional view of genes as beads along the chromosome necklace to the more modern idea of the *fluid genome*.

RNA SYNTHESIS

Can DNA, then, itself form the 'mould' or 'template' for protein synthesis for which we are searching? Several pieces of evidence make this unlikely. Almost all the DNA of the cells is in the chromosomes, which are themselves, at least in animal and plant cells, contained within the nucleus. (Some DNA is also present in mitochondria and, in plants, in chloroplasts, and the significance of this will become apparent in Chapter 14.) In some organisms it is possible to remove the nucleus, and hence all the DNA, by dissection. The enucleate cells of one alga, called acetabularia, can continue to live for some weeks after the operation has been performed, and during this time will continue to synthesize proteins quite as efficiently as the whole cell. Similarly, it is possible to make preparations from cell homogenates which are free of DNA, and which nonetheless will incorporate radioactive amino acids into new protein. Even treating such homogenates with an enzyme (*DNase*), which will hydrolyse and thus remove any last traces of DNA, will not prevent protein synthesis. So we must conclude that, although DNA is the substance transmitted during cell division and reproduction, and hence forms 'the' genetic material ultimately responsible for carrying the information relevant to protein synthesis, it does not form the immediate template on which the protein is made. This function is reserved for the other of the two nucleic acids, RNA, and the flow of genetic information in normal cells runs from DNA to RNA and then to protein, a process considered so fundamentally irreversible by Francis Crick that he once called it the 'Central Dogma' of molecular biology (Figure 30).

Like many such dogmatic statements, it can be misleading and is now known to be not entirely true (amongst many other such processes, a class of viruses called *retroviruses*, of which HIV, the AIDS-producing virus, is but one example, can reverse the flow from DNA to RNA).

RNA, it will be recalled from Chapter 3, like DNA, is a chain composed of purine and pyrimidine bases, in this case adenine, guanine, cytosine, and uracil. Unlike DNA, the sugar component is not deoxyribose but ribose itself. And, also unlike DNA, its molecules are not composed of two strands intertwined into a helix, but of single chains. There are three major forms in which RNA is found in the cell which have diffferent roles in the information transfer process from DNA to protein synthesis. *Messenger* or mRNA contains the information required for the amino acid sequence of the proteins. The relatively low molecular weight RNAs that select and carry amino acids to the place of protein synthesis are called *transfer* or tRNA. Finally, the bulk exists, as was described in Chapter 4, in tiny insoluble particles, distributed in the cytoplasm and

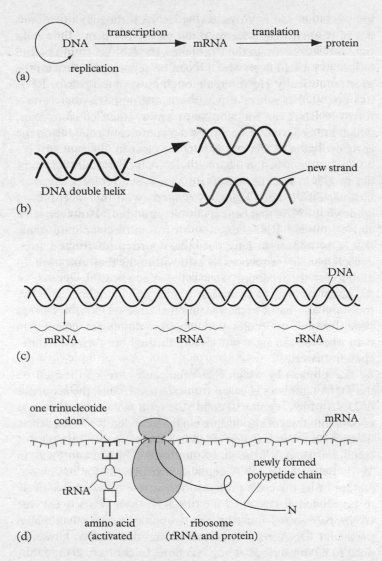

Figure 30. The role of DNA and RNA in nucleic acid and protein synthesis. (a) The central dogma. (b) Replication. (c) Transcription. (d) Translation.

known as ribosomes. *Ribosomal* or rRNA is the major component of ribosomes and forms about ninety per cent of the cell's
total RNA content. In the ribosomes, the RNA is found bound
tightly to protein from which it can be separated only with the
greatest difficulty. However, although most of the cellular RNA
has one of these three functions, it also appears that certain
forms of RNA can function as enzymes, so-called ribozymes,
which are believed to have played a significant role during the
early evolution of life, as will become clear in Chapter 14.

As was described in Chapter 3, RNA is chemically very similar to DNA, with the exception that uracil replaces thymine.
Cellular RNAs are all single-stranded (with the exception of
some viral RNAs that can be double-stranded). However, such
single-stranded RNAs often form intramolecular base pairs,
thus they appear to have secondary structures formed from
complementary sequences located within the molecule itself. In
addition to the secondary structures, made possible because of
the large number of hydrogen bonds they contain, the RNA
molecule also has a tertiary structure. Unlike DNA, RNAs are
very dynamic molecules that undergo significant changes in
conformation during synthesis and throughout their entire lifespan in the cell.

The process by which RNA molecules are synthesized on
the DNA templates is called *transcription*. In this, the sequence
of DNA which is going to define the template for the protein is
copied into the corresponding single-stranded RNA sequence.
The process was first described in a classic paper by François
Jacob and Jacques Monod, of the Pasteur Institute in Paris, in
1961. Because the DNA contains instructions for the manufacture of all the cell's proteins, only a few of which are likely
to be needed at any one time, the RNA-synthesizing enzyme
(*RNA polymerase*) must be able to recognize or select which
particular DNA regions, representing the desired proteins,
need to be transcribed at any one time. In contrast to the DNA
polymerase, RNA polymerase does not need a primer.

Transcription starts at specific sites, called *promoters*, on the
DNA template. Sequencing the DNA of the large number of

transcription start regions has shown that certain sequences occur with great regularity. These so-called consensus sequences can act as molecular switches regulating the synthesis of an RNA molecule. The promoter region of the DNA is thus *not* itself part of the message eventually transcribed into RNA; by contrast those regions of the DNA which are transcribed into RNA and eventually copied into protein are referred to as representing *structural* genes.

How does the enzyme locate the promoter sites? Just as with DNA replication, transcription starts with chain opening, a conformational change in the DNA molecule that forces the hydrogen bonds between the bases apart. The RNA polymerase then moves along the DNA template strand in the $3 \rightarrow 5$ direction, linking the matched RNA bases so as to make an 'antiparallel' copy of the DNA in which the RNA will be an exact copy not of the template DNA but of the opposing DNA strand. (Another difference of course is that because RNA contains uracil instead of thymine, it is this uracil that pairs with the DNA adenine base.) The RNA polymerase moves along the DNA double helix until an entire 'transcript' length of RNA has been produced, and the transcribed region of DNA regains its double-helical conformation as the next section of DNA unwinds. The newly synthesized chain is almost immediately released from the DNA, thus allowing synthesis of many RNA copies of the same gene in a relatively short time. Transcription lacks the proofreading activity that is observed during replication. As a consequence, RNA polymerase makes about one mistake for every 10^4 nucleotides, in comparison to DNA polymerase which has a much lower error rate of about one in 10^7 nucleotides.

This mechanism enables a length of RNA to be produced as a copy of the genetic material, the DNA, which corresponds to the protein which is ultimately required. The overall reaction rate is much slower than that of replication, only about 40 nucleotides per second. As the average gene is about 1500 nucleotide pairs long, it will be completely transcribed in about 90–100 seconds. The process is shown in Figure 31.

*Figure 31. The transcription of a DNA base sequence into an mRNA
base sequence*

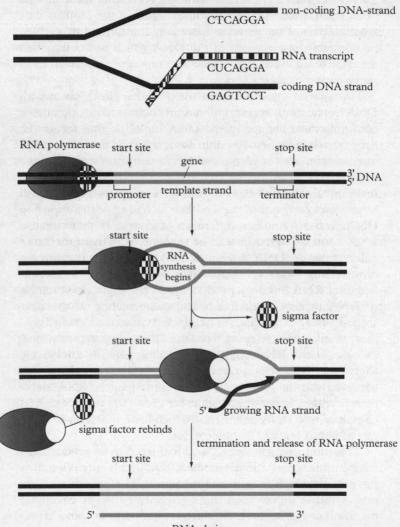

RNA polymerase was originally defined simply by its ability to catalyse the synthesis of RNA under the direction of a DNA template. But polymerase has some other characteristics that make possible the precise initiation and termination of RNA synthesis. RNA polymerase is a large multisubunit complex that is able to associate with different polypeptides needed to recognize initiation or termination signals at particular genes. The best-characterized RNA polymerases are those of bacteria, such as the common gut bug *E. coli*, on which much of this research has been done. Although there are some differences in the way in which RNA polymerases recognize the transcription start sites in bacteria and in eukaryotes, the major principles are very similar. The complete enzyme can be separated into two components, the *core enzyme* (made of four subunits) and the *sigma factor*, which is required to initiate precise transcription. The sigma factor is released when the RNA chain is 8–9 bases long, causing significant change in the conformation of the enzyme. When the growing chain extends to 15–20 bases, the core enzyme makes a further transition to form the complex that undertakes elongation.

The termination of transcription is as precisely controlled as its initiation. The DNA template contains stop signals which are still not fully known. However, a common feature is that a GC-rich region before the termination site is followed by an AT-rich sequence. The most distinctive feature of termination sequences is the symmetrical repetition of this GC-rich region, and that the RNA transcript of this region is itself 'self-complementary' – which means that its sequence of bases can pair with each other to form a hairpin structure, as shown in Figure 32.

In addition, the growing RNA chain ends with several uracil residues specified by the series of adenines in the AT-rich region of the DNA template. One or more of these structural features causes RNA polymerase to pause when it encounters them; it can then be released from the DNA. In some cases, a specific protein called a *rho protein* is essential for chain termination. It is thought that rho, a protein made of four identical

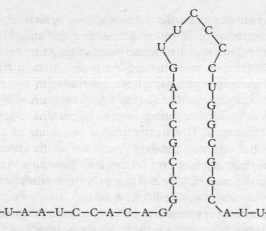

Figure 32. Base sequence of C3 end of an mRNA transcript (the tryptophan operon) from the bacterium E. coli. *A stable hairpin structure can be formed*

subunits and with a molecular weight of around 200 000, attaches to RNA and moves towards the RNA polymerase at the termination site, where it displaces the enzyme from the C3 end of the RNA, thus releasing the transcribed mRNA and allowing the DNA double helix to re-form. This releasing process, too, is ATP-dependent.

The transcription process in eukaryotes is more complex than in prokaryotes because eukaryotic RNA polymerase consists of three distinct enzyme forms capable of transcribing only a single class of cellular RNA. mRNA synthesis is catalysed by the RNA polymerase II, tRNA by RNA polymerase III and rRNA and some other small RNAs synthesis is carried out by RNA polymerase I. In addition, chromatin must be made accessible to all RNA polymerases, additional transcription factors must bind to the promoter region for a gene to be active, and other sequences, located some distance away from the promoter region, called *enhancers*, must bind protein factors to stimulate transcription.

As the DNA on which the RNA is synthesized lies in the nucleus (with the exception of a small amount in the mitochondria and chloroplasts, the significance of which is discussed in Chapter 14) and proteins are made in the cytoplasm, the RNA molecule that is the immediate product of transcription must leave the nucleus through the pores in the nuclear membrane mentioned in Chapter 4 and visible in Plate 4. However, before it does so, the so-called *primary transcript* undergoes several processing steps depending on the type of RNA being produced. The future mRNA molecule will be modified at the 3'-end by polyadenylation, that is, the addition of a 3'-terminal poly(A) tail, and on the 5'-end by the addition of methylated guanine. It is believed that these two modifications increase the stability of mRNA.

It has become apparent that the genes whose DNA has been sequenced (and this is now a rapidly increasing number) are not in fact arrayed in simple linear sequence. Instead, the DNA sequences which make up the gene are separated by long stretches of DNA which appears not to be transcribed at all. These sections of DNA, interrupted genes, have become known as *intervening sequences* or *introns*, whilst the sections of the DNA that seem to be genetically relevant are known as *exons*. Even that DNA which is transcribed into RNA seems to be present in excess so far as the requirements of protein synthesis are concerned. Thus in multicellular organisms much more RNA is copied from the DNA than will ever finally be released from the nucleus.

The discrepancy between the organization and the length of the gene and the organization and the length of its RNA is a result of the process known as splicing. The process is shown diagrammatically in Figure 33. Splicing of the primary transcripts occurs in the nucleus, together with the other modifications which are mentioned in the case of mRNA. Splicing accomplishes the removal of introns and the joining of exons into the mature RNA which constitutes the message which is finally conveyed to the ribosome.

Nuclear splicing requires the formation of a *splicosome*, a large particle that assembles the primary transcript, and small nuclear ribonucleoproteins, snRNPs, called, unattractively,

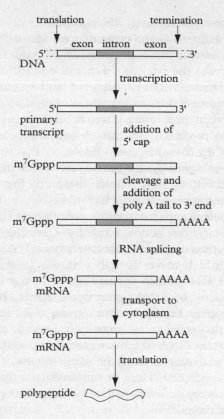

Figure 33. Different stages in eukaryotic gene transcription

snurps (no publisher would have accepted *that* term!). Snurps are composed of a single uridine-rich RNA and several proteins. Splicing can be broadly divided into two stages: in the first, the sequence at the border between exons and introns is recognized by the RNA in the snurps and, in the second, cleavage and ligation reactions bring two separated exons together. Enzymes are not involved in the processing of mRNA – however, they are required for tRNA splicing. It is the mRNA molecule itself that carries out the entire reaction of splicing. A group of snurps assembles at each intron on the primary transcript, and first cuts the RNA at the 5' splice site, separating the left exon. In the second stage, a cut at the 3' splice site releases the free intron and the exon on the right side is ligated to the exon on the left side. The released intron is rapidly degraded (Figure 34). Splice sites are not specific for individual RNA precursors. However, the RNAs of some genes are capable of alternative splicing which brings together different exons from the same gene, thus allowing different proteins to be produced by the same gene.

Such observations however in no way answer the question of what all that intron DNA is doing in the cell. Indeed, molecular biologists remain somewhat baffled by the question, as is indicated by the variety of names that it has been given – repetitive DNA, junk DNA, even selfish DNA. Some have concluded that it has simply been accumulated during evolution by replication, and once it is there there is no way of getting rid of it because it will automatically replicate whenever a cell divides or reproduces – hence Crick's name for it of 'selfish' DNA; others, notably François Jacob, have suggested that it might provide a useful store of potentially valuable sequences for evolutionary processes to 'tinker' with. And indeed, the non-coding sequences do contain regulatory sequences that are recognized by a class of proteins called regulatory *transcription factors*, discussed in more detail in the next chapter. Cells have extensive mechanisms of gene regulation that allow them to respond to changes in their environment by altering the expression of specific genes. These sequences are often hidden in non-coding regions of the DNA. It is certainly counter-intuitive to believe that there should be such a large macromolecular fraction, so energetically expensive to produce, yet with no clear function in the economy of the cell.

It serves to remind us how little we still know about many of the fundamental mechanisms involved in those two questions of genetic interest which we posed at the beginning of this section – the origins of similarities and the origins of differences. All the cells in any individual organism's body contain the same DNA sequences, yet express very different proteins. Equally,

Figure 34. Excision of introns

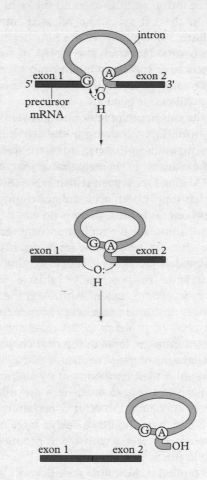

two closely related species, like chimpanzee and human, carry some 98.5 per cent of their exon DNA in common – yet no one would mistake a chimpanzee phenotype for a human one, and at the present time we have no idea of the reasons, locked within the biochemistry and the regulatory mechanisms of development, why we and the chimpanzee differ as much as we do.

FROM RNA TO PROTEIN

We must now leave the mRNA, poised as it is to leave the cloistered environment of the nucleus for the hurly-burly of cytoplasmic society, and return to the question of how the amino acids to be turned into protein are prepared in their turn for attachment to the mRNA template. The first step in this preparation consists of the activation of the amino acids, so as to prime them with sufficient energy for the subsequent formation of peptide bonds. This activation is performed at the expense of ATP by a series of *amino acyl-tRNA-synthase* enzymes, one for each amino acid. In their presence, amino acids react with ATP to form amino-acyl-AMP complexes:

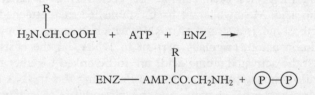

In the next stage of preparing the amino acid for the template the activated amino acid is transferred from the enzyme complex to a transfer tRNA molecule, of which there are several alternative ones for each amino acid. Many tRNAs have been purified, and their structures determined – indeed tRNAs, which are around 80 nucleotides in length, were the first nucleic acids to be sequenced. The amino acyl synthase enzyme catalyses the formation of a covalent bond between the carboxyl group of the amino acid and the free hydroxyl group at the C3 terminal end of the tRNA molecule, releasing the synthase enzyme once more. All tRNA molecules have been found to have the same trinucleotide sequence of cytosine-cytosine-adenine (CCA) at the end which links to the amino acid, and they all seem to be folded into the same secondary structure, rather like a four-leaved clover (Figure 35). The molecule thus has three other arms besides the one that becomes charged with the amino acid. These other arms of the clover leaf are characterized by the presence of certain unusual nucleotides, and each has a specific function. One of the arms, for example, is responsible for recognizing the ribosome to which the tRNA is shortly to become attached.

To recapitulate the story to date, we have arrived at a situation where each amino acid about to become incorporated into protein is now linked to a specific tRNA. To prepare the amino acids for this incorporation, the cell has already had to provide more than twenty different enzymes, one for each amino acid, and twenty different tRNAs, as well as two phosphate bonds per activated amino acid. It has also been necessary not only that the enzymes be specific for individual amino acids, but that a new class of specific molecules is introduced – a set of tRNA molecules each capable of 'recognizing' a different amino acid. Meanwhile, we have established that, attached to the ribosome, is a strand of mRNA which contains a copy of the information, originally present in DNA, on the basis of which the activated amino acids are to be linked together in a specific primary protein sequence. Providing this mRNA has also involved other enzymes and the expenditure of energy.

We now come to the key aspect of the entire process. There has to be a relationship between the amino acids and the sequence of nucleotide bases along the mRNA chain, which is adequate to specify the sequence in which the tRNA linked amino acids will form up.

The RNA molecule consists of 4 bases arranged along a chain. How can a sequential structure of 4 different units be arranged so as to 'recognize' a set of 20 different units? The answer to this question was provided in the mid 1950s by the physicist George Gamow. He argued that, if each base corresponded to a different amino acid, only 4 different amino acids could be recognized by the chain of bases. If it needed a combination of 2 bases to code for each amino acid, 16 different amino acids could be recognized, as there are 16 (or 4×4)

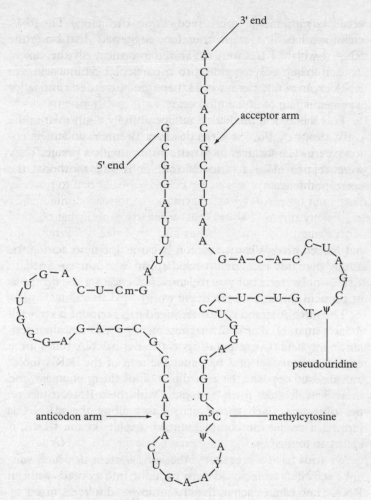

Figure 35. Clover leaf structure of yeast tRNA for phenylalanine

different possible ways of combining the four different types of base into groups of two. Similarly, if 3 bases corresponded to each amino acid, 64 (or $4 \times 4 \times 4$) different amino acids could be recognized (Table 6). As there are up to 20 different amino acids found in proteins, the minimum number of bases that could correspond to each 'recognition site' along the RNA chain must be 3. Gamow therefore suggested that, along the RNA template, bases were organized into successive groups of 3, each group corresponding to a particular amino acid. An RNA chain of 600 bases would therefore be needed to code for a protein chain of 200 amino acids.

This theoretical prediction was beautifully confirmed in the early 1960s by Francis Crick, Sidney Brenner, and their coworkers in Cambridge, in genetic studies with a phage. They were able to make a series of mutations and show that the correspondence:

$$3 \text{ bases} \equiv 1 \text{ amino acid}$$

did indeed hold. What is more, it became apparent during the 1960s that this recognition mechanism was not confined to viruses or bacteria but was the universal code for the synthesis of proteins throughout the living world.

The mRNA strand is thus arranged in a series of units, each 3 bases long, each of which represents, or codes for, a particular amino acid. Corresponding to these 3 mRNA bases are a complementary set of 3 bases on the arm of the tRNA molecule directly opposite the arm which binds the amino acid, and these 3 nucleotides form base-pairs with the mRNA triplet on the ribosome. Each triplet of 3 bases along the mRNA is termed a *codon*; the complementary triplet on the tRNA is called an *anti-codon*.

We thus have a protein-synthesizing system in which suitably activated amino acids are brought into contact with an RNA chain where each individual amino acid is 'recognized' by a group of three bases and held in position by them. When enough of the amino acids required to match all the groups of bases on the RNA chain are present, the peptide bonds between adjacent amino acids can be formed, the activating tRNA molecule released into solution, and the new protein molecule unpeeled from its RNA template.

Is there any way of telling which triplets of bases correspond to which amino acids? The answer to this question was spectacularly provided by Marshall Nirenberg and Heinrich Matthaei of the US National Institutes of Health, Bethesda, Maryland in 1961. Nirenberg and Matthaei argued that if a synthetic RNA chain made up of only one type of base could be made, such a chain ought to provide a template for the synthesis of peptide chains containing only one type of amino acid. By using an RNA synthesizing enzyme, *polynucleotide phosphorylase*, they were able to make an artificial RNA chain containing only the base uracil (polyuridylic acid, which can be represented … UUU …). When they tried the effect of this in a protein-synthesizing system, in which the artificial RNA chain replaced the natural mRNA, they discovered that in its presence a peptide chain containing only one sort of amino acid – phenylalanine – was made. Thus the sequence of bases UUU is the RNA recognition site for the amino acid phenylalanine. This breakthrough was rapidly followed up by both Nirenberg and Matthaei and by Severo Ochoa in New York. RNA polymers made up of differing combinations of the four bases yielded differing artificial peptide chains, and by the end of 1966 virtually the entire RNA code language for amino acids was known. Polynucleotide phosphorylase was indeed the Rosetta stone of protein synthesis and it is appropriate that the process of synthesizing protein according to a defined RNA sequence is known as *translation*.

The relationships between nueleotide triplet 'codons' and the amino acids for which they code is shown in Table 6. The sixty-four possible combinations of the four RNA bases into groups of three is required to code for only twenty amino acids, and there is therefore some *redundancy* amongst the codons; three different groups, UCU, UCC and UCG, all code for the same amino acid, serine, for instance. In the parlance of information science, the RNA code for proteins is 'degenerate'. Four of the sixty-four codons, though, do not represent amino acids at all, but are signals which mean, for instance, 'start reading' and 'stop reading' – that is, they indicate the beginning and end of a protein chain, just as a capital letter and a full stop

Table 6

	AGA								
	AGG								
GCA	CGA						GGA		
GCC	CGC						GGC		AUA
GCG	CGG	GAC	AAC	UGC	GAA	CAA	GGG	CAC	AUC
GCU	CGU	GAU	AAU	UGU	GAG	CAG	GGU	CAU	AUU
Ala	Arg	Asp	Asn	Cys	Glu	Gln	Gly	His	Ile
A	R	D	N	C	E	O	G	H	I

imply the beginning and end of a sentence. The 'full-stop' codons are UAG, UAA and UGA. They designate chain termination, and are not read by tRNA molecules but rather by specific proteins called *release factors*.

The start signal for protein synthesis is more complex. In bacteria, polypeptide chains start with a modified amino acid, *formylmethionine*. (The situation in eukaryotes is slightly different.) There is a specific tRNA for formylmethionine, which recognizes the codon AUG (or less frequently GUG). However, AUG is also the codon for an *internal* methionine residue, and GUG the code for internal valine. This means that the signal for the first amino acid in a polypeptide chain must be more complex than for all subsequent ones. AUG (or GUG) is *part* of the initiation signal, and a further signal, preceding the codon, determines whether it is to be read as chain initiation signal or a codon for an internal residue.

Let us look more closely at the molecular events at the ribosome, where the recognition and formation of peptide bonds actually occur. As mentioned in Chapter 4, the ribosome is composed of two subunits, one much smaller than the other. The initiation of protein synthesis is a complex process that must bring together the small ribosomal subunit, the aminoacyl-tRNA complexes and mRNA, all in an appropriate orientation. Initiation also requires the presence of an energy source (supplied not by ATP but GTP), a particular amino acyl tRNA

Leu	Lys	Met	Phe	Pro	Ser	Thr	Trp	Tyr	Val	stop
UUA					AGC					
UUG					AGU					
CUA				CCA	UCA	ACA			GUA	
CUC				CCC	UCC	ACC			GUC	UAA
CUG	AAA		UUC	CCG	UCG	ACG		UAC	GUG	UAG
CUU	AAG	AUG	UUU	CCU	CCU	ACU	UGG	UAU	GUU	UGA
Leu	**Lys**	**Met**	**Phe**	**Pro**	**Ser**	**Thr**	**Trp**	**Tyr**	**Val**	**stop**
L	K	M	F	P	S	T	W	Y	V	

(that is amino acyl formylmethionine in bacteria) whose anticodon corresponds to the 'start here' codon on mRNA and, at least in bacteria, a number of soluble protein 'initiation factors' (abbreviated as IF) which bind transiently to the ribosome. In the first step IF2 binds to GTP and Met-tRNA. Initiation now proceeds with the binding of small ribosomal subunit and IF3, which in turn binds a further set of initiation factors. This large complex is 'scanned' for the correctness of the location of the AUG codon on the small ribosomal subunit, an ATP-requiring process. Only then can the formation of the initiation complex proceed and the large ribosomal subunit can be attached, with addition of IF5.

The ribosome has three sites for tRNA binding: the P site (peptidyl site), the A site (aminoacyl site), and the E site (exit site). Initiator tRNA, the one carrying methionine, is the only tRNA that can bind to the P site – all other incoming amino acyl tRNAs bind to the A site. In other words, an amino-acyl-tRNA is placed in the ribosomal A site near the P site already occupied by methionyl-tRNA. Now *peptidyl transferase*, an enzyme which forms an integral part of ribosome, catalyses the first peptide bond formation between the amino-group of amino-acyl-tRNA and the carbonyl carbon of the methionyl-tRNA. Formation of this peptide bond will result in the transfer of the methionine to the amino group of the amino-acyl-tRNA in the A site. The deacylated tRNA, previously loaded with methionine, can now be transferred from the P site to the E site. In the next step a translocase enzyme moves the mRNA and tRNA carrying dipeptide from the A site to the vacant P site. The energy for this movement is provided by GTP. Now, the ribosome, with a vacant A site, is ready for the new cycle – to accept the third amino acid to be incorporated into the growing polypeptide chain.

The next step is *elongation* of the polypeptide chain. As in almost all processes described until now, the fidelity of elongation is enhanced by the presence of another family of protein factors, elongation factors (EF), which aid the process of selection of correct amino-acyl-tRNAs at the expense of GTP. One of these factors, together with amino-acyl-tRNA and GTP, binds to the ribosome and if the hydrogen bonding indicates that the right amino acid is in place, the amino-acyl-tRNA will be placed at the A site. As a consequence, GTP will be hydrolysed and the GDP-EF complex released. The protein chain will grow in a particular direction, so that the first amino acid of the sequence has a free —NH2 group whilst the last amino acid has a free —COOH. Successive cycles of peptide bond formation and translocation of the growing polypeptide chain on the ribosome, mediated by the elongation factors, will continue until the *chain terminating codon* is reached. UAG, UAA or UGA codons will not allow RNA to bind. Instead, they bind a protein called a release factor (RF), in the form of a complex with GTP and the entire polypeptide chain is released. Protein synthesis stops and the newborn protein is released, at the expense of hydrolysis of GTP.

The mechanism is shown diagrammatically in the sequence of Figure 36. It can best be compared to the functioning of a tape recorder in which the ribosome acts as the pick-up head moving steadily past the mRNA tape, reading the RNA code as it goes and converting the code words into the language of proteins. Like a tape recorder, the ribosome is neutral as to the message it plays; this is provided by the mRNA tape. The ribosomal machinery is available for whatever mRNA tape (or even artificial RNA, as in the Nirenberg and Matthaei experiments) appears on the scene.

Granted that the mRNA is simply a long sequence of bases, it is important that no mistakes are made in reading it. A sequence AUGUUUCAGACC, for instance, *could* in principle be read in triplets starting with the first A, or that first A could be considered part of a *preceding* triplet, or indeed, it could be part of more than one set of overlapping messages. It was a matter therefore of some relief to those studying protein synthesis when it became apparent that in most circumstances there is only one way to read the mRNA message, as a *nonoverlapping* triplet sequence. Thus the sequence of bases

AUGUUUCAGACC

would be read	AUG	UUU	CAG	ACC
and translated as	Met	Phe	Glun	Thr

The exact role of the many different ribosomal protein and RNA components in ribosomal function is not fully understood. Obviously the physical stability of the ribosome is of prime importance to protein synthesis and it is in the maintenance of this stability that Mg^{2+} and K^+ ions play a part by ensuring an ionic environment which prevents dissociation of the ribosomal structure.

This read-out system makes it possible for more than one portion of the long template molecule to be read, by different ribosomes, at a given time and, in fact, in cells which are actively synthesizing proteins, it is possible to isolate by suitably gentle techniques, such template RNA molecules with several ribosomes attached to different portions of them. Such a cluster of ribosomes held together by a template is called a *polysome* (Chapter 4, Figure 37; also Plate 4). This simultaneous read-out enables many molecules of the same protein to be made at the same time in response to cellular demand.

Figure 36. (a) Initiation of protein synthesis.

(b) Elongation of protein synthesis.

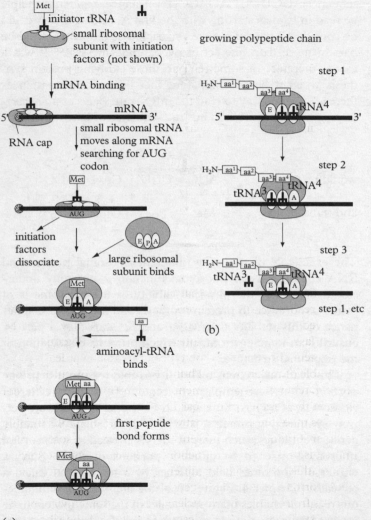

Figure 36. (c) Termination of protein sythesis

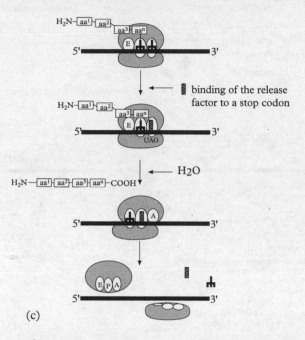

(c)

It is apparent from this that each mRNA molecule can be used to read out a protein several times over (just as a tape can be played many times on a tape recorder). How long the mRNA lasts varies from minutes to days, depending on the cell and protein concerned. Some types of template, at least, must be capable of making several hundred protein molecules before needing renewal, as experiments demonstrating protein synthesis in the absence of nuclear DNA reveal.

It was mentioned above that whilst most of the DNA in multicellular organisms is present in the nucleus, some small amount is present in the mitochondria (and chloroplasts); what is more, this DNA, which is different in sequence from nuclear DNA, forms a unique set of genes, coding for a different set of proteins from those of the nuclear DNA. Although virtually all protein synthesis occurs as described so far, on the ribosomes

Figure 37. Schematic drawing of polysome

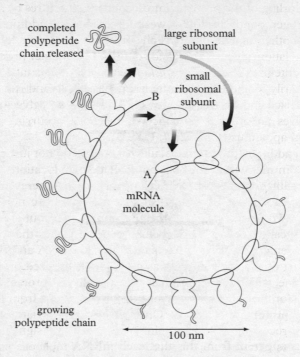

completed polypeptide chain released

large ribosomal subunit

small ribosomal subunit

B

A

mRNA molecule

growing polypeptide chain

100 nm

present in the cytoplasm, the mitochondria contain their own protein-synthesizing system, including the necessary RNA, and can and do also make certain proteins including, interestingly enough, some, but by no means all, of the proteins required for their own activity as ATP-synthesizing structures.

The story of protein synthesis is not concluded, of course, by the production of the polypeptide chains spun off from the mRNA by way of the ribosomes. The reading of the DNA sequence by way of transcription and translation processes provides a mechanism for the extremely accurate copying of the chains, so accurate that many millions of polypeptides are likely to be synthesized before a single error is likely to crop up by chance. Nonetheless, many are not yet finished proteins; they represent only a set of primary structures. To turn them into proteins requires the linking of subunits by disulphide bridges, the curling of the proteins into secondary structures by way of hydrogen bonds and other weak forces, and the addition of all those other moieties that give the protein its character, by phosphorylation, glycosylation and so forth. This further processing of proteins, called *post-translational modification,* takes place primarily in and on the endoplasmic reticulum and is a highly organized and specific activity. Thus, for instance, glycosylation involves an ordered sequence of reactions occurring in the Golgi apparatus (Chapter 4; Plate 3).

In addition, protein molecules, of course, do not just exist in random mix within the cell, but have definite locations, in the subcellular organelles, at the cell membrane, or wherever. In recent years it has become clear that molecules are not left to their own devices to get to the appropriate sites, but are rather specifically labelled. The labelling generally takes the form of extra lengths of polypeptide, coded in the mRNA and present as part of the protein chain as it is first synthesized on the ribosome, which serve as signals that a particular protein is destined for the cell membrane, or possibly for export from the cell as a hormone or whatever. Once at their appropriate site further processing of the proteins takes place, the signalling sequences are chopped off and the finished, smoothly fitting molecules inserted into place.

Envoi: Despite abundant remaining areas of ignorance, both practical and theoretical, the period since the Watson–Crick paper of 1953 has undoubtedly proved the most intellectually exciting in the entire history of biology in the extent to which Crick's 'Central Dogma' of molecular biology has become understood but now transcended. And, as will become clear in the final chapter, based on these experiments, theories and techniques, biochemistry and molecular biology are now moving rapidly from mere intellectual excitement into a full flood of production in the burgeoning industries fuelled by biotechnology. There are some molecular biologists who would claim that the 'golden age' is now past, that most of the vital issues are now resolved, and there is nothing left to do but fill in the odd pieces of the jigsaw. For others, the best is yet to come, for they believe – and they are very articulate in defending their claim – not merely that all of biochemistry, but eventually, probably sooner rather than later, physiology and even ecology, will be 'reduced' to the inexorable working out of genetic programmes.

It is true that the analysis of cellular properties not merely in terms of energy but also of information flow is essential, and this will become more apparent as we move to the questions of control of metabolism discussed in the next chapter. Nonetheless, as this chapter has hopefully made clear, to argue that all of biochemistry can be subsumed under the unilateral flow of information DNA $\rightarrow$ RNA $\rightarrow$ protein disregards (a) the fluidity of the genome and the variety of ways exons can be combined, (b) the complexity of RNA processing in the nucleus, (c) which proteins are expressed at any time from a particular DNA, (d) how those proteins function in the cell, and (e) how they are modified and modulated by the environment of the cell itself and the past history of that cell. These issues will become more apparent in the chapter that follows. To give primacy to any one component, even so exciting a one as the construction of primary protein sequences, is to miss the interactions that bind them all. It is to these that we should now turn.

CHAPTER 11

Controlling the Cell

The last chapters have presented the cell to us almost as if it could be thought of as a highly organized industrial plant. We have pictured raw materials, such as amino acids, entering through the cell membrane and being processed on a production line of ribosomes geared to produce proteins based on information derived from DNA in a sort of nuclear planning office. These proteins are then sent to other parts of the cell to themselves perform further operations on fresh raw material. Meanwhile, the cell's fuel, glucose, is pulped down to pyruvic acid and dispatched to the mitochondrial boiler-house, where it is converted to useful energy which is distributed throughout the cell and used to drive the entire machinery forward. Such metaphors of the cell as machine are revealing; certainly biochemists and molecular biologists do tend to see the cell in this way; there has historically been a tendency in biology to derive analogies for living systems from technological artefacts and to use them almost without recognizing that they are *only* metaphors, after all. Metaphors can be helpful but one should beware of pushing them too far; the logic of living systems is more subtle than that of any human-built machine yet devised, and it sometimes might be preferable to reverse the analogy and think of machines as if they were organisms!

In any event, in the elegant picture of the cell as a piece of interlocking machinery directed by DNA something is seriously missing. We have emphasized the integrity of the cell as a system, that each part needs all the others in order to survive. Without ATP, no protein synthesis; without proteins, no ATP. Yet, to continue the machine metaphor for a moment, industrial processes, however efficiently organized, cannot indefinitely operate smoothly and without attention. Crises and emergencies arise, production lines become fouled, raw materials are erratic in arriving and half-finished goods may accumulate.

Changes of plan are frequently required to readjust output to the demands of a fluctuating situation. All these changes, in response to both internal and external pressures, require that decisions are constantly made, decisions which must be based on information about activities everywhere within the factory, and on the ability to assess transient conditions. And if so comparatively crude a system as a factory needs its decision-making system, one might argue, how much more so must the delicate and sensitive mechanism of the cell?

The control and regulation of the cell, not only in its day-to-day running, but in its entire cycle from birth to death within the body, and indeed the entire cycle from life to death of the body itself, can be interpreted biochemically, provided always we are prepared to extend our understanding of biochemistry above the level of individual molecular interactions to embrace the structural properties of the cell as a whole, as a *system*. It is increasingly possible to understand how the complex regulatory processes, that in the past appeared bafflingly to be the product almost of conscious will, are really no more and no less than the inevitable consequences of the combination of the appropriate organization of physical structures and chemical reaction sequences that occur within the cell, following from them according to principles which can often be given mathematical form.

We can begin by examining the factors that regulate a simple enzymic reaction. Think of a biochemical reaction

$$X \underset{\text{X-ase}}{\rightleftharpoons} Y \qquad (1)$$

catalysed by the enzyme X-ase. We can write a long list of conditions which affect the rate at which X is converted into Y. They include the concentrations of X and Y, for, assuming the reaction is fully reversible, the substance that exists in the highest concentration at any time will force the reaction to proceed in the direction of its conversion to the other. This is the law of mass action. The amount of X-ase present will affect the rate too, because if there is just a small amount of enzyme it will soon become saturated with its substrate and will become the sole limiting factor; a further increase in substrate concentration will have no effect on the reaction (page 116). We have also seen how temperature can affect the rate of the reaction; if it is too hot the enzyme will even be destroyed completely by denaturation. A change in the ionic concentration surrounding the enzyme molecule can have a drastic effect on its activity because of the need to maintain the correct electrical charges at the active centre. Many enzymes too, as we have seen, require the presence of various cofactors before they can function, and of course enzymes – especially allosteric ones – are very sensitive to the presence of activators or inhibitors whose binding causes a conformational change that alters the shape of the active site. Finally, the thermodynamic equilibrium of the reaction is also important: it may lie in the direction of complete conversion of X to Y; of almost no conversion; or at some point midway between the two.

Any or all of these factors will affect the rate of the reaction; that is, the number of molecules of X converted into Y every minute, although in general it is found that change in any particular one of the variables has a much greater effect than changes in the others. For example, the effects of small changes in temperature and pH may be negligible, and all reactants may be present in optimal concentrations, but X-ase may be obligatorily dependent on magnesium as an activator. Under such circumstances, very slight alterations in the amount of magnesium present may have very large effects on the rate of production of Y.

Now consider the variables affecting a reaction pathway, say the conversion of W to Z by way of the intermediates X and Y:

$$W \rightleftharpoons X \rightleftharpoons Y \rightleftharpoons Z \qquad (2)$$
W-ase X-ase Y-ase

For each of the three reactions involved, the set of variables we have described will control the reaction rate. But, over the entire reaction sequence, an additional factor will now operate that did not exist when only one reaction was being discussed, because the rate of each reaction is also affected by the outcome of the others. Suppose, for example, that the W-ase reaction is extremely sensitive to changes in pH, and that the final reaction product, Z, is an acidic substance. As more and more Z is produced, the acidity of the solution will increase. But when this happens, the enzyme W-ase will be affected and the rate of production of X from W slowed. As the amount of X produced declines, so will the amounts of Y and Z. Thus the Z content will cease to rise and the acidity of the solution will decrease. Immediately, the rate of W-ase will once more be accelerated, Z and acidity will rise, and W-ase activity will decline. Ultimately, a steady state will be reached.

Thus the production of Z exerts a controlling influence on W-ase, and, reciprocally, the rate of W-ase controls the production of Z. This type of control is called *feedback*. It is identical in type to the principle of a thermostat on an electric heater, or the governor of James Watt's steam engine 200 years ago. We can represent a reaction sequence with feedback like this

$$\text{W} \; \underset{\text{W-ase}}{\rightleftharpoons} \; \text{X} \; \underset{\text{X-ase}}{\rightleftharpoons} \; \text{Y} \; \underset{\text{Y-ase}}{\rightleftharpoons} \; \text{Z} + (\text{H}^+) \qquad (3)$$

Feedback can be of two types. In the example above, Z production *inhibited* W-ase – this is *negative feedback*. But if Z production lowered the pH so that W-ase was accelerated and Z production increased, this would be *positive feedback*. In all feedback reactions, the principle is the same – the rate of a reaction is controlled by a substance which is an ultimate product but not itself directly involved in the reaction.

When we come to consider the thousands of reactions and hundreds of reaction sequences that occur within the cell, all of them controlled by many different variables, and many themselves altering these variables by generating acidity or alkalinity, using or producing cofactors and inhibitors, the immediate feeling may well be one of despair. With so much going on, how can one sort out just what is, and what is not, significant?

But though it certainly remains complex, one can simplify the problem. Consider again the reaction sequence:

$$W \rightleftharpoons X \rightleftharpoons Y \rightleftharpoons Z$$
$$\text{W-ase} \quad \text{X-ase} \quad \text{Y-ase}$$

If we ask what is the *maximum rate*, under the most favourable circumstances, of each of the enzymic reactions involved, and then express these rates in arbitrary units, we may find, say:

$$W \rightarrow X = 100$$
$$X \rightarrow Y = 10 \qquad (4)$$
$$Y \rightarrow Z = 1$$

What this means is that the W-ase reaction is ten times as fast as the X-ase reaction, and one hundred times as fast as the Y-ase reaction. What is now the *rate-limiting* step in the production of Z from W? Clearly W-ase is producing X at ten times the rate that X-ase can use it, whilst X-ase is also producing Y at ten times the rate that Y-ase can use it. As both X and Y are thus being produced far in excess of the amounts which can be handled by their respective enzymes, the rate of conversion of W to Z is limited only by the rate of the Y-ase reaction, $Y \rightarrow Z$. Thus it would seem that the rate of a reaction sequence is controlled by the rate of the individual reactions of that sequence. Under these circumstances, even if Z production results in a change in pH big enough to cause a ten-fold decline in the rate of W-ase, the rate of Z production will remain unaltered, as this depends not on W-ase but on Y-ase. Z will no longer be exerting feedback control over its own production.

Thus, if all we are considering is a set of enzymes in the sequence and a starting substrate then the only reaction which would matter, so far as control is concerned, is the slowest reaction of the sequence, and the rate of any given sequence of reactions is regulated by altering the variables which control the rate of just one of those reactions – the slowest. This is where a study of enzyme kinetics and a knowledge of parameters such as Vm, as described in Chapter 5, can be helpful. We would thus expect to find in every reaction sequence one or more critical control points and we may also expect to find them at an early reaction in the pathway, as this would prevent the futile use of enzymes further along the pathway and the build-up of useless intermediates. Similarly, we could expect such control points to occur very close to the branching point of pathways that serve more than one function, so that one branch could be regulated without affecting the others.

However, this neat linear picture is still rather too simple, because it only takes account of one reaction sequence at a time, as if it were indeed isolated from everything else going on in the cell simultaneously. In reality there are many thousand enzyme-catalysed reactions occurring, some relatively independently, but most linked together in a complex web of interconnections – indeed the term *metabolic web* has been used to describe them. Under these circumstances, it can be shown mathematically, using either algebra or, more recently, the fashionable chaos theory, that stability and control are vested in the entire web, rather than any one reaction sequence within it. When cellular reaction rates are studied in more or less intact preparations, such as tissue cultures or slices of the sort described in Chapter 6, what limits and controls the rates of reaction sequences like glycolysis includes not one simple rate-limiting step but many of the factors that we must now turn to consider.

INTERNAL REGULATORS OF METABOLISM

Control of energy production

The bulk of the energy demands of the cell is met within the mitochondria by the production of ATP during the oxidation of substrates by way of hydrogen/electron transport (see Chapter 7). When the enzymes and carriers of this system are studied in isolation, they are capable of extremely rapid reactions, yet if the intact mitochondrion is presented with substrates such as pyruvic acid it is found that the rate of pyruvic acid oxidation reaches a maximum which is considerably below the maximum velocities shown by the individual carriers. As increasing the amount of pyruvic acid does not alter this oxidation rate, it is clear that the mitochondrion must contain its own built-in control system to limit the rate at which it burns fuel. We can isolate some of the elements in this control system by drawing a schematic flowsheet of the operations involved in oxidation (Figure 38).

From this diagram it can be seen that the rate of oxidation of substrate will be held up if:

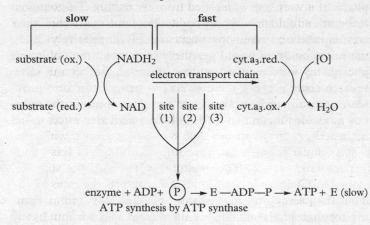

Figure 38. Flow-sheet for oxidative phosphorylation

(a) not enough oxygen is present to cope with the production of reduced cyt. a_3 at the end of the transport chain. But normally, unless poisons such as cyanide or carbon monoxide are present, the cell has plenty of oxygen available and the oxidation of cyt. a_3 does not represent a rate-limiting step;

(b) not enough inorganic phosphate and ADP are present. This may not be immediately obvious, but consider what would happen if either ADP or phosphate were not present. Energy trapped within the proton gradient across the membrane could not be tapped off for ATP production. The ATP synthase which forms the chemiosmotic pump would grind to a halt, and reduced intermediates of the electron transport chain would accumulate, leading to a build-up of the first reduced hydrogen carrier ($NADH_2$). When the levels of ADP and phosphate within the mitochondria rise again, the pump can begin work and oxidation proceed. It therefore follows that the rate of oxidation of substrate through the electron transport chain is controlled by the concentrations of ADP and inorganic phosphate present in the mitochondrion.

However, the concentrations of ADP and inorganic phosphate are themselves controlled by other factors. The terminal reaction of oxidative phosphorylation tends to lower their concentration by removing them as ATP. Thus, left to itself, the mitochondrion would gradually remove all the ADP and phosphate, and cease to oxidize substrate. But at the same time, in other parts of the cell, ATP is being utilized to provide energy for protein synthesis, muscular contraction, nervous transmission, and so forth. All these activities result in the net reaction

$$ATP \longrightarrow ADP + \textcircled{P}_i \tag{5}$$

Thus the energy-utilizing reactions *increase* ADP and inorganic phosphate levels and hence tend to *speed up* oxidation. In any given cell system, the rate of substrate oxidation will therefore be dependent on the rate of ATP utilization. Thus the cell can adjust its rate of energy production in order to meet the varying demands upon it. The critical link that gives this adaptability is provided by the ratio of ATP to (ADP+ $\textcircled{P}_i$) within the cell, a ratio critical to the control of energy metabolism. It is the gear that links the turning of the cell engine to the revolution of its wheels.

Like all gears, it can be slipped into neutral if oxidation through the electron transport chain can in some way be uncoupled from phosphorylation. This can be done by inducing a conformational change in the coupling factor, the ATP synthase that is driven by the chemiosmotic proton pump. The catalytic activity of this enzyme is dependent on the strongly hydrophobic environment provided by the membrane. A conformational change in the enzyme molecule will expose the active site to water on the edge of the membrane, so it can then act as an ATPase, hydrolysing ATP to ADP and inorganic phosphate. So temporary uncoupling will decrease the ATP/ADP ratio until it reaches a level when the concentration of ADP will no longer be inhibitory and oxidation can once more be linked to phosphorylation. Several drugs, antibiotics and poisons possess this uncoupling power, as do several naturally occurring substances within the cell.

Some ATP and ADP molecules are always present in the cell, and the ratio between them is an important controlling factor not merely within the mitochondrion. Thus phospho-fructokinase, which phosphorylates fructose-6-phosphate to the more reactive fructose 1, 6, diphosphate (page 164) in the early stages of glycolysis, is an allosteric enzyme which is activated by the binding of ADP and inhibited by binding ATP.

If ADP, inorganic phosphate and oxygen are all present in abundance, none of the control mechanisms we have just outlined will be rate-limiting for oxidative phosphorylation. Under these circumstances yet another reaction becomes the limiting factor, the *entry* of hydrogen into the transport sequence by way of the dehydrogenase reactions:

$$AH_2 \rightleftharpoons A + NADH_2 \tag{6}$$

Like all reversible reactions, the direction in which the dehydrogenase reaction is driven will depend on the relative concentrations of the various products and substrates present. The reaction will thus go forward only when the amounts of AH_2 and NAD are large and those of A and $NADH_2$ small. As there are many dehydrogenase reactions supplying hydrogen to the transport chain, all having NAD as a common cofactor, the rate of all of them will depend on the relative amounts of NAD and $NADH_2$ present. When NAD is abundant and the reduced $NADH_2$ only present in low quantities, all the dehydrogenase reactions will tend to move forward, the tricarboxylic acid and fatty acid oxidation cycles that link them will begin to revolve, and a steady stream of hydrogen atoms will be dispatched down the carrier line to oxygen. But if $NADH_2$ begins to accumulate and NAD diminish, the dehydrogenase reactions will slow down to a halt and hydrogen transport cease.

Now, consider what happens if a muscle cell, say, steadily producing energy and oxidizing substrate, is called upon for a sudden burst of activity. ATP levels are reduced, ADP levels rise, and the transport chain accelerates. As it accelerates, $NADH_2$ is reoxidized more rapidly and the ratio of NAD to $NADH_2$ increases. As this ratio increases, the dehydrogenase reactions of the oxidation cycles are pushed forward more rapidly and the substrates of oxidation burn yet faster as the cycles spin round. The cell thus provides more energy in response to demand in the same way as the car engine revolves more rapidly at the touch of the accelerator. As the demand for energy slackens, ATP and $NADH_2$ accumulate, hydrogen transport slows down, and the oxidation cycles reduce to a tickover rate only.

There is another interesting side to the controlling role played by the $NAD/NADH_2$ ratio in energy metabolism. We have already referred to the fact that there exist two closely related coenzymes, NAD and NADP, both of which can be alternately oxidized and reduced. NAD is the coenzyme involved in the cell's oxidative, energy-providing systems, whilst NADP finds a place in the synthetic processes, such as fat synthesis, which requires a steady supply of hydrogen for reducing reactions. For these reactions, $NADPH_2$ is demanded, and, just as the ratio of NAD to $NADH_2$ is a control point for oxidation, so the $NADP/NADPH_2$ ratio is a rate-limiting factor for synthetic reactions.

We can now appreciate the rationale behind the specificity of usage of the two coenzymes; if only one of them was used in both types of reaction it would be impossible to control anabolism and catabolism independently of each other by this method of altering the ratios. An even finer control is exerted by enzymes called *transhydrogenases* which catalyse the reversible reaction

$$NADH_2 + NADP \rightleftharpoons NAD + NADPH_2 \qquad (7)$$

This reaction thus simultaneously oxidizes one coenzyme and reduces the second. By so doing, it serves to alter simultaneously and in reverse directions the ratios $NAD/NADH_2$ and $NADP/NADPH_2$. By altering these two ratios, which between them control the rates of the oxidative and reductive pathways of the cell, transhydrogenase helps to hold the balance between the catabolic, energy-yielding reactions on the one hand and the anabolic, energy-requiring reactions on the other.

Control of biosynthetic pathways

Discussion of the transhydrogenases has moved us from oxidative to synthetic pathways. Of the many biosynthetic pathways operating in all cells there has been space here to discuss only a few: lipid, glycogen, and protein synthesis. We have scarcely referred at all to the means whereby the cell fabricates the precursors of these big molecules – amino acids and purine and pyrimidine bases for example. Yet for the synthesis of each of these there is a chain of enzymes which builds complex from simpler molecules. Each chain will have its own specific rate-limiting steps and feedback mechanisms, many of which have now been analysed in detail, particularly in organisms such as bacteria which are anyhow more biosynthetically versatile than animals.

Although each biosynthetic system shows certain unique features, there are some general mechanisms that seem to apply. One such was provided by Edwin Umbarger, in New York in 1956. He was studying the biosynthesis in microorganisms of the amino acids lysine, methionine, and threonine from glucose by way of a simpler amino acid, aspartate, and its metabolite aspartate-semialdehyde. When the bacteria were fed radioactive glucose, radioactivity was subsequently found in the three amino acids. But when the organisms were grown in a medium containing, for instance, unlabelled threonine, then incorporation of radioactivity into this particular amino acid was abolished. The presence of the amino acid thus resulted in a suppression of its continued synthesis – a perfect example of negative feedback. This process, called *end-product inhibition*, means that the biosynthesis of a given amino acid is regulated by the amount of the amino acid itself being produced as the end-product of a reaction sequence. As this amount increases, so further biosynthesis is reduced; if its concentration falls, synthesis starts up again. This form of feedback control has since been found to be widespread in biosynthetic systems, in animals as well as microorganisms, and even within the context of the metabolic web is probably a major regulatory mechanism. Its most interesting feature can be illustrated if the pathway is drawn as follows:

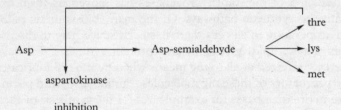

Umbarger was able to show that suppression of synthesis of threonine was caused by the threonine inhibiting the first enzyme along its biosynthetic pathway, i.e. aspartokinase. However, production of the other amino acids was unaffected, even though they too are made from the same precursor, aspartate-semialdehyde. This is because there is not simply one aspartokinase enzyme but three, each with essentially the same enzymic activity but with a slightly different protein structure. Enzymes which exist in multiple forms like this are now known to be fairly frequent. They can generally be separated by electrophoresis as they differ slightly in ionic charge, and are known as *isoenzymes*. The existence of multiple isoenzyme forms provides another powerful regulatory mechanism. In the case of aspartokinase, each of the three isoenzymes is specifically inhibited by one of the three amino acids, threonine, lysine and methionine.

All three must be inhibited before production of the intermediate ceases completely. In addition, each of the amino acids inhibits the first enzyme on its branch line away from aspartate-semialdehyde, so ensuring that the decrease in concentration of the metabolite affects only the production of the inhibiting amino acid. This is an example of control by enzyme multiplicity, although the individual inhibitions are brought about by the type of allosteric processes that have already been described.

A control mechanism that achieves the same effect as this, although by a different mechanism, is the multi-end-product inhibition of allosteric enzymes. Glutamine synthase (responsible for the production of glutamine from glutamic acid) is regulated by this mechanism. It has binding sites for trytophan, histidine, and ATP among others, because glutamine acts as an amino-group donor in the biosynthesis of all these compounds. If only one of these locks into the synthase then glutamine production will be turned down by a small amount. If two of them bind, then production will be lowered a little more. And so on. The effect is thus progressive, until all the end-products have been bound, when glutamine production ceases altogether.

REGULATION OF GENE EXPRESSION

An even more ingenious method of control was first identified amongst microorganisms by Jacques Monod and François Jacob in Paris in the 1960s, work which was fundamental to present-day understanding of genetic control. They showed that many microorganisms, which normally lacked the enzymes to deal with particular substrates, would, if grown in a medium containing these substrates, at once begin to synthesize the missing enzymes. In particular, Monod and Jacob studied bacteria lacking in the ability to metabolize the sugar galactose. Presented with galactose as substrate, which would normally be metabolized by way of galactose phosphate and UDP-galactose, the bacteria of one strain (though not of others, where even the power of synthesizing the galactose enzymes had been lost) proceeded to fabricate the enzymes needed to utilize it. The new enzymes that were required were synthesized completely from scratch – that is, a sequence of novel protein synthesis was initiated. This enzyme synthesis could be triggered off even if galactose itself was replaced by a non-metabolizable substance which resembled it closely in structure. As the enzymes began to be made immediately in response to the substrate, the genetic information necessary for their synthesis must have been present in the cells all along. But the proteins were not normally present. How could this intriguing effect be explained?

Monod and Jacob worked with *E. coli*, which normally lacks the ability to metabolize the sugar lactose (β-galactoside). When *E. coli* is grown in the absence of lactose the enzyme which metabolizes it, β-galactosidase, is not required, and indeed there is very little of it present in the cells – no more than about 5 molecules per bacterium. However, bacteria need to respond quickly to changes in their environment, particularly as they often cannot choose the nature of the medium in which they find themselves, arriving more or less by chance. So challenged by the need to metabolize lactose, they set about synthesizing the enzymes required to do so. Thus within 2–3 minutes of adding lactose to the medium in which the bugs are growing, enzyme activity increase rapidly to about 5000 molecules per bacterium. Remove the substrate, and the synthesis of the enzyme stops as rapidly as it began. Moreover, the bacteria increases the activity not merely of the β-galactosidase, but of two other enzymes, permease, required to help the sugar substrate across the cell membrane, and transacetylase, which transfers an acetyl group from acetylCoA to β-galactosides. How could this intriguing effect happen?

It turns out that bacterial genes are often organized into clusters that comprise a set coding for proteins with related functions. Such clusters are known, following Monod and Jacob, as *operons*. Such operons include both so-called *structural* genes – that is genes that code for particular proteins, as described in the previous chapter, and a number of other genes whose function is to regulate the activity (or *expression*) of the structural genes. These *regulatory* or control genes include both *inducers* and *repressors*. A regulatory gene codes for a protein (or RNA) which alters the expression of other genes. Inducers and repressors are small molecules which start (induction) or stop (repression) the expression of particular genes. It is an inducer which switches on the synthesis of the β-galactosidase, and a repressor which turns it off again.

Monod and Jacob termed the gene cluster controlling the β-galactosidase synthesis the *lac operon* and argued that it occurred because, in the absence of the unusual substrate, the synthesis of the enzymes required to metabolize it was repressed, thus ensuring that the cell did not waste time or energy fabricating enzymes for a biosynthetic pathway it would not require. In the presence of the substrate, enzyme synthesis was induced by the added galactose or related substances.

If the mRNA coding for the new enzymes is already present, then induction of their synthesis can occur by stimulating its translation. If it is not present, however, it would be necessary to make new mRNA molecules by transcription from DNA. Bacterial mRNA, unlike the mRNA of animals, has a very short life, half of it decaying within about three minutes, so it seems unlikely that synthesis could be controlled by regulating translation. If new mRNA molecules were made, however, then transcription of the new mRNA could continue as long as the inducer was present. In the bacteria studied by Monod and Jacob, the genes controlling the synthesis of all the enzymes required for galactose metabolism (their structural genes) are situated along a consecutive length of DNA, enabling them to be controlled as a group. A little way off is another gene responsible for the manufacture of a specific repressor protein. We have seen that mRNA is transcribed by a polymerase enzyme that binds first of all to a DNA promoter site. In the absence of the inducer, the repressor protein, according to the Jacob and Monod hypothesis, binds to a site on the DNA between the promoter and the structural genes and thus prevents the RNA polymerase from moving along from the promoter to transcribe the DNA. The site to which the repressor protein attaches is called the *operator* region. The inducer exerts its effect by specifically combining with the repressor molecules so that they can no longer bind to the operator and RNA polymerase is unobstructed so that transcription can now occur (Figure 39).

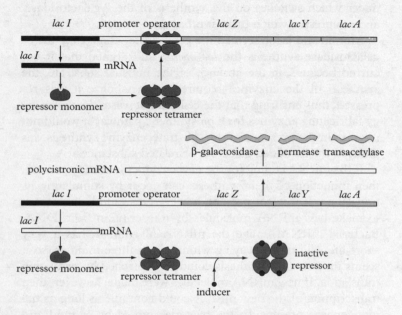

Figure 39. The lac operon

A cell which can control its protein synthesis so as to make only the enzymes immediately demanded by its metabolic situation – even though it still retains the genetic DNA codes carrying the coding sequences for the manufacture of others, should this become worthwhile because of an abnormal diet or unusual situation – is clearly highly adaptable and well qualified to survive extreme variations in conditions. It is not surprising therefore that regulatory systems based on the operon principle have been found to be widespread among microorganisms.

CONTROL OF GENE EXPRESSION IN
MULTICELLULAR ORGANISMS

The challenges that multicellular organisms face are very different from those of bacteria. Firstly, such organisms can choose their environment much more readily than can bacteria, and thus search out food materials they are able to metabolize rather than having always to be able to adapt to novelty. Furthermore, because it is the task of the organism as a whole to preserve the relative constancy of its internal environment, on homeodynamic lines, the range of conditions which individual tissue cells have to meet is fairly tightly controlled, at least in health. The price that has to be paid is however that each cell loses something of its individuality, and instead becomes specialized as a liver or brain, or leaf or root cell. Brains have a different set of enzymes from livers and roots from leaves, and yet of course all the cells of any organism contain a set of all the genes that organism possesses. So there have to be means whereby one set of proteins is expressed in one cell type, and another in another. And this isn't the end of the complications. The same cell type may express one set of proteins during early development and a different or more restricted range later in life. And finally, in a multicellular organism, where the life of any cell is subordinate to the life of the whole, there must be ways by which cells in one part of the body can communicate with those in others, signalling them to change the rate of production of some molecule or other. These signals are provided by nerves, discussed in the next chapter, and by substances circulating through the bloodstream in animals, or diffusing through intracellular space in plants, which can interact with particular cells and thus regulate their activity. Such signalling substances are called hormones, whose mode of action we turn to later in this chapter.

However, before we come to consider these further complexities, we should focus on the ways in which eukaryotic protein synthesis is regulated at the level of the nucleus. The mammalian genome contains about a thousand times more

DNA than does *E. coli*, and, as discussed in Chapter 10, the coding sequences are often interrupted by intervening, non-coding sequences, the introns. Eukaryotic DNA is associated with the histones and the nucleo-histone complex prevents transcriptional access to much of the DNA most of the time. This seems to be due to the histones stabilizing the DNA in such a way that makes chain separation by breaking of H-bonds more difficult. Further, in multicellular organisms, where the DNA is largely confined to the nucleus, mRNA is not simply attached to the ribosome and transcribed, but, as we saw in Chapter 10, is subject to a great deal of further processing in the nucleus before being permitted to leave by way of the nuclear pores. Sections of mRNA corresponding to particular exons are spliced together, introns are excised and considerable further processing occurs. The consequence is that most of the RNA synthesized in the nuclei of multicellular organisms is degraded before it ever enters the cytoplasm, thus providing a further source of control.

Transcription in eukaryotic cells depends upon three different RNA polymerases present in the nucleus, polymerase I, for rRNA, II for mRNA, and III for tRNA and some other small RNAs. Eukaryotic RNA polymerase II cannot initiate transcription unaided, but requires the presence of a group of proteins called *transcription factors* (TFs) at the promoter before transcription can begin. Another difference from bacteria is that the regulatory proteins don't have to be so close to the promoter but can be thousands of base pairs away. Thus a single promoter can be controlled by a large number of regulatory units scattered along the DNA. Most such control is positive, inducing rather than repressing synthesis (Figure 40).

Most eukaryotic promoters carry a short DNA sequence, primarily of T and A nucleotides, and thus called a TATA box, located about 25 nucleotides upstream of the start point of transcription. The first step in initiating transcription is the formation of a complex containing the TATA box, and a binding protein, transcription factor IID, which together position the polymerase and enable it to bind to the promoter. A variety of other transcription factors then become engaged, which first serve to shape the DNA into a form which makes copying possible, and then enable the polymerase to move away from the promoter and begin the formation of the phosphodiester bonds which elongate the growing RNA molecule. At this point most of the TFs are released from the DNA.

As if this were not complex enough, the activity of eukaryotic promoters is often further controlled by the presence of a further group of sequences located some distance away from the promoter itself – indeed on either side of the gene – called *enhancer* sequences. This additional twist means both that different proteins can regulate the expression of a single gene, and also that a single regulatory protein can coordinate the expression of many different genes. Particular proteins and, as we will see, hormone-receptor complexes, can bind to DNA sequences called *response elements*, occurring upstream of promoters and enhancers, which makes possible the further control of the expression of regulatory genes (Figure 41).

The enormous possibilities of control offered by the presence of enhancers, response elements and transcription factors, and the hormonal signalling mechanisms they interact with help regulate the huge range of proteins required in the many different cell types of the multicellular organism – some 250 in humans for instance – in the great trajectory of life from conception through infancy and maturity to old age. Protein synthesis is regulated at many levels: by specific gene activation, the initiation of transcription, itself controlled by interacting transcription factors, processing the transcript, transport to the cytoplasm and translation of the RNA. There is a whole class of genes – so-called immediate early genes (IEGs) whose function when activated seems to be to synthesize proteins which then regulate the expression of other ('late') genes. It is the proteins coded for by these late genes which do most of the business of the cell which the previous chapters have discussed. Nothing is simple in the life of the cell, and the straightforward Central Dogma by which 'DNA makes RNA makes protein' is increasingly recognized as a brilliant simplification of the complex metabolic orchestration which constitutes our inner life.

Nowhere is this clearer than in the context of development. All multicellular organisms begin as fertilized eggs, containing, as we have made clear, the entire complement of genes for all the many different cell types in the body. As the single starting cell divides, it gives rise to daughter cells whose progeny in their turn will in due course have many different developmental fates, to end as liver or muscle or hormone secretors, as roots, branches or leaves. Each cell type is, as we have already mentioned, characterized by a specific pattern of gene expression. It is the proteins regulating transcription which perform such an important role in the regulation of development. The systematic manner in which these regulators are turned on and off helps ensure that eyes become eyes and brains brains.

Control by structure

Chapter 4 described how the cell is built up of a series of subcellular organelles, nucleus, mitochondria, ribosomes and others, all embedded in the cytoplasm which is bounded by the cell membrane. Each organelle contains characteristic enzymes and substrates, and those present in one organelle are not necessarily freely available to those in another: the control of release of mRNA from nucleus into cytoplasm is one example; another is offered by the dangerous hydrolytic enzymes, which, as described in Chapter 4, are kept firmly caged within the lysosomes, and allowed out only under extreme conditions. Thus the cell can be seen as consisting of a series of separate 'compartments' bounded by membranes which control what passes across them. Environmental conditions inside and outside an organelle may be very different; there may be differences of acidity and alkalinity, temperature, concentration of ATP or essential cofactors such as magnesium. It should not be assumed that, because the overall concentration of a substance within the cell is high, it is present in the same high concentration in all parts. Some regions may have none at all. And even apparently slight differences may have formidable consequences. A temperature difference across a mitochondrial

Figure 40. Initiation of transcription of a eukaryotic gene – different phases of formation of initiation complex

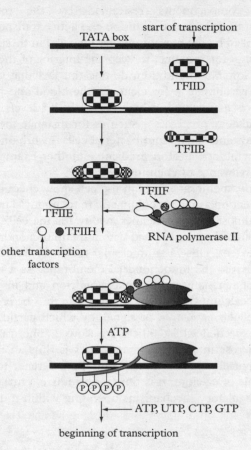

membrane of only 0.01°C does not sound much, but so thin is the membrane that this represents a temperature *gradient* of ten thousand degrees per centimetre. Under such conditions many hitherto unsuspected control systems may function.

What *is* known is that cellular membranes are extremely fussy about what they will and will not allow to pass through them; as has been discussed in Chapter 4, even quite small molecules may have their free diffusion into and out of the cell and its compartments restrained by the membrane. Mechanisms of facilitated diffusion and active transport, which are controlled by the membrane, are involved in these processes – helping, for instance, to keep the interior of the cell in a high K^+, low Na^+ condition, despite the fact that the extra-cellular environment – for example, the blood and plasma in animals – is maintained at high Na^+, low K^+ levels. Many of these regulatory processes – such as for instance the entry of amino acids and sugars into bacterial cells – work on the basis of transmembrane proton gradients, another example of the universal relevance of chemiosmosis.

Similarly, membranes within the cell show chemical selectivity. The endoplasmic reticulum and mitochondrial membrane have an almost vacuum-cleaner avidity for the cell's calcium, which is sucked up and stored within the mitochondrion to an amazing extent. This vacuum cleaner too is driven by a proton gradient across the mitochondrial membrane. As a very large number of reactions in the mitochondrion and in the cytoplasm are calcium-dependent, calcium can thus be regarded as an intracellular *messenger*, or signal, by which particular reactions are speeded whilst others are slowed. The transport of calcium across intercellular membranes is thus an important cellular control mechanism. But with this reference to the regulatory role of calcium, it is time to introduce a further major class of regulatory mechanisms operating within multicellular organisms.

EXTERNAL REGULATION OF CELL METABOLISM: THE HORMONES

As well as functioning as an internally self-regulating, homeodynamic mechanism, as has already been emphasized, cells in a multicellular organism are part of a larger whole. Each is one of many hundreds of millions in a particular organ, the organ one of a group of interdependent parts of the body. The

Figure 41. (a) Different regions of a eukaryotic gene. (b) Looping of
the DNA brings into contact multiple regulatory elements

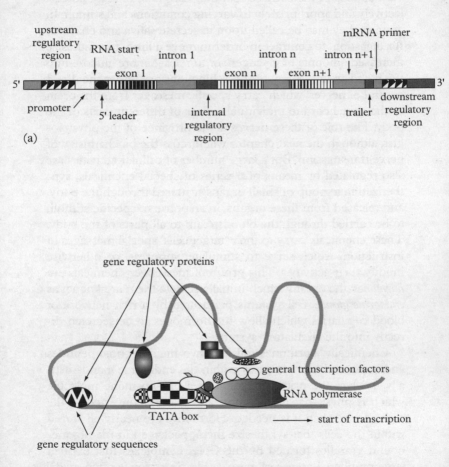

survival of the cell is not the ultimate goal; rather the function of the cell is subordinated to the survival of the organism, and, by reproduction, of the species. The organism, then, must have means of interlocking and coordinating the functions of the various cells of the body so that they can unite to respond collectively and appropriately to varying conditions and stimuli. In animals they may be called upon to secrete saliva and enzymes for digestion, to contract in order to move a limb, to synthesize increased amounts of glycogen, or to release more glucose into the bloodstream. Many of these functions are under the direct control of nerves, which carry specific messages from the brain or the spinal cord to individual organs of different parts of the body. The role of these nerves is the province of the physiologist, although the next chapter will discuss the biochemistry of nerve transmission. But a large number of cellular activities are also regulated by means of a series of special chemicals, synthesized in a group of small organs scattered through the body, and released from these organs, in response to specific stimuli, to be carried through the bloodstream to all parts of the body. These chemicals carry to their 'target cells' special messages of instruction, which serve to stimulate, suppress, or otherwise modify their activities. This group of messenger-chemicals are *hormones*; the organs which initially secrete them are known as *endocrine glands*, cell systems permeated by a rich network of blood capillaries which allow the hormones to be secreted directly into the circulatory system.

Chemically, hormones fall into two main groups, peptides and steroids. Their synthesis within the endocrine glands usually involves the prior formation of a pro-hormone molecule which is physiologically inactive and is only converted to the active form when it is needed. Hormones are nearly all stored within the cells that synthesize them, packaged in membrane-bound granules formed by the Golgi complexes that exist in large numbers in these secretory cells. Secretion is often mediated by other hormones or by nervous control and the circulating levels of several of the hormones fluctuate over particular time periods (for instance, daily, monthly) with precise regularity. They are nearly all transported in the blood complexed with plasma proteins, whose function is to protect the hormone from degradative attack until it is safely at its target; the protein-hormone complexes also serve as a reservoir for the hormone as they are in equilibrium with free hormone. As the level of free hormone drops so more hormone is dissociated from its protein carrier to maintain the equilibrium. Hormones thus enable the organism to exert subtle control over a host of metabolic functions and to alter these functions in response to changes in conditions in the outside world.

Circulating in the bloodstream of the human there may be as many as fifty different chemicals, which, between them, are capable of causing substantial differences to the metabolic pattern of the target cells which respond to them. We cannot here describe all the hormones or all their functions. In many cases they are extremely complex, but all hormonal activity can ultimately be related to the effect of the hormone in affecting the rate of some intracellular metabolic sequence. However, hormones are present only in low concentration even in the bloodstream and are never found free within the cytoplasm or nucleus. Instead, their interaction with the target cell is mainly by way of highly specific receptors, either on the cell surface or within the cytoplasm; the bound hormone-receptor complexes then activate a series of further intracellular reactions ranging from redistribution of ions like Ca^{2+} across membranes to gene expression.

The results of a slight change in the balance of cell metabolism may differ profoundly for the animal depending on whether the cell concerned is, say, in the liver, the brain, or the testis. Although the physiologist may find in one case the onset of fatty liver, in another coma and mental aberrations, in a third abnormal sexual development, the biochemical mechanics behind all three may be virtually identical. Physiologically, the sex hormones are involved in the onset of puberty and menopause, ovulation, menstruation, pregnancy, parturition and milk production, and secondary sexual characteristics, to say nothing of their interaction with the brain resulting in involvement in complex changes in behaviour and motivation. Yet to the bio-chemist none of these more interesting features is relevant (at least during working hours) and here we shall concentrate on the general cellular processes involved in their activity.

There are a number of ways of grouping hormones. The simplest is according to the glands that secrete them. Thus we have *thyroxine*, produced by the thyroid gland in the neck, and *parathormone*, made in the tiny parathyroids which lie at each side of the thyroid. *Insulin* and *glucagon* are secreted by the pancreas. Above each kidney there is a small gland called the adrenal, the central region of which produces *adrenalin* whilst the outer region (the cortex) produces a battery of *steroid hormones: aldosterone, corticosterone*, etc. The testis and ovary produce the sex hormones, *testosterone* and *oestrogen* (though neither is exclusively present in one sex; their names are distinctly misleading in this regard). Finally, a minuscule organ, in humans the size of a pea, located in a cavity of bone between the bottom of the brain and the palate at the top of the mouth, produces twenty-five or more different hormones, which have the subtle task of regulating the output of virtually all the other hormones in the body. This control gland is the *pituitary*. For our purposes, however, a better classification is provided by considering the distinct biochemical principles by which different hormones act, and on this basis they can be divided into two classes: those, generally *peptides*, which do not enter their target cell but act by interacting with a receptor on the cell membrane, and those, like the *steroids*, which also interact with cytoplasmic receptors or transcription factors.

The basis of hormone action: second messengers

The binding of hormones to receptors on the cell membrane is analogous to the binding of enzymes to substrates or pros-thetic groups. Many of the techniques for receptor study are already familiar from earlier chapters. They include labelling the receptor with a radioactive molecule similar in structure to the hormone that will bind to it. Such a substance is termed a *ligand* and may even evoke the same physiological response as does the hormone, in which case it is called an *agonist*. *Antagonists* are substances that bind to the receptor but do not evoke a similar physiological response and are widely used as inhibitors of hormone action. Radioactive labelling makes it possible to determine the kinetics of the ligand-binding reaction and to show that binding is usually very rapid but reversible.

Nearly all the receptors that have been characterized have turned out to be glycoproteins, and this helps explain perhaps their most important and dramatic property – their ability to discriminate between very similar ligands and to specifically bind the one hormone that structurally interacts with the binding site. Kinetic studies show that any one receptor molecule is unlikely to bind more than a single hormone molecule. But these small numbers of interactions somehow trigger an enormous and often long-lasting response within the cell. This led to the conclusion that the hormone signal must be *amplified* in some way by the cell. One way of performing this amplification would be for the hormone *(first messenger)* to trigger the production of another, and it is this second substance *(second messenger)* that then affects the intracellular target system.

In the late 1950s, Earl Sutherland and his co-workers in Baltimore, studying the ways in which the hormones adrenalin (an amine) and glucagon (a polypeptide) speeded up glycogen breakdown to glucose in the liver, succeeded in isolating a compound that, when administered to slices of target tissues, was capable of mimicking not merely their effects, but those of other hormones too. This compound was identified as a form of the nucleotide adenosine monophosphate (AMP) in which the molecule is bent back on itself as a ring, *cyclic AMP* or *cAMP*.

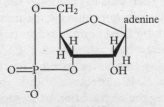

Cyclic AMP

Cyclic AMP is formed from ATP by the action of *adenylyl cyclase*, a membrane-bound enzyme, and is destroyed by a specific *phosphodiesterase* enzyme which hydrolyses it to AMP. In the absence of phosphodiesterase, cAMP is a very stable compound. Sutherland's work led to the concept that cyclic AMP is a second, *intracellular messenger* in the action of many hormones, the first messenger being the hormone itself. The essential features of this concept are:

1. Cells possess receptors for hormones in their external membranes.

2. The combination of a hormone with its specific receptor in the cell membrane leads to a stimulation of adenylyl cyclase, which is also bound to the membrane.

3. The increased activity of adenylyl cyclase increases the amount of cAMP inside the cell.

4. The cAMP then acts inside the cell to alter the rate of one or more functions.

An essential feature of the second messenger model is thus that the hormone need not enter the cell. It has its effect at the cell membrane and then its biological effects are mediated inside the cell by cAMP rather than by the hormone itself. The evidence for this includes the observation that adenylyl cyclase within a target cell is stimulated only by hormones that affect that cell; hormones which do not elicit a biological response in a particular cell do not affect the adenylyl cyclase activity of that cell. It has also been proved that the concentration of cAMP in target cells changes in direct proportion to the biological response of such cells to hormonal stimulation. Further, the biological effects of a hormone can be mimicked by the addition of cAMP or a related compound to target cells.

Such experiments have shown that cAMP is a second messenger for many hormones in addition to glucagon and adrenalin. Cyclic AMP affects a huge range of cellular processes, including the degradation of storage fuel molecules, the secretion of hydrochloric acid by gastric mucosa, the dispersion of melanin pigment granules, and the aggregation of blood platelets.

How does the binding of a hormone such as adrenalin or glucagon to a specific receptor lead to the activation of adenylyl cyclase? The hormone-binding sites lie on the extracellular surface of the cell membrane whereas the catalytic sites of adenylyl cyclase lie on the intracellular surface. More importantly, hormone-binding sites and adenylyl cyclase catalyic sites are located on different proteins, which can be separated by centrifuging a detergent-treated preparation of the cell membrane. Adenylyl cyclase is a large membrane protein of molecular weight 185 000. The receptor for adrenalin has a molecular weight of 75 000 and is also known as β-adrenergic receptor because as well as adrenalin it binds a spectrum of pharmacologically active compounds.

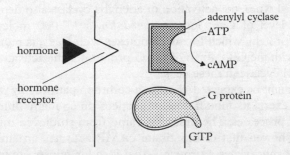

Figure 42. The activation of adenylyl cyclase by the binding of a hormone to its specific receptor

Adenylyl cyclase is not directly activated by the binding of hormone to a specific receptor. The effect is mediated by a third protein, called a *G protein* because it binds GTP and GDP. This regulatory protein (molecular weight 42 000) exists in two forms, the GTP complex which activates adenylyl cyclase and the GDP form which does not. Inactive GDP is converted into active GTP by the exchange of free GTP for bound GDP. This GTP–GDP exchange is catalysed by the hormone-receptor complex but not by the receptor alone. Thus, the signalling system implies an information flow from the hormone receptor to the G protein and then to adenylyl cyclase, as illustrated in Figure 42.

The G protein has yet another property which enables it to function as the information-carrying intermediate between hormone receptors and adenylyl cyclase. GTP bound to the G protein is slowly hydrolysed to GDP; thus the G protein functions as a GTP-ase. This provides the regulatory protein with a built-in mechanism for deactivation. The proportion of G protein in the GTP state, and hence of adenylyl cyclase in the active form, depends on the rate of exchange of GTP for GDP compared with the rate of hydrolysis of bound GTP. The rate of GTP–GDP exchange is greatly enhanced by the binding of G protein to occupied hormone receptors. Consequently, nearly all of the G protein is in the GDP form and nearly all the adenylyl cyclase inactive when the hormone level is low. In some cell types, the activation of adenylyl cyclase also depends on the level of Ca^{2+} ions. *Calmodulin*, a 17 000 molecular weight protein, which forms complexes with Ca^{2+}, is required along with the GTP form of the G protein for the activation of adenylyl cyclase in these cells.

As might be expected, there is a common pathway by which cAMP exerts its multifarious influences on so many different cellular processes. The first clue came from studies in muscle which showed that in this tissue cAMP activates an enzyme that phosphorylates a protein – that is, adds phosphate to the serine or threonine residues of the amino acid chain (Chapter 3). Such phosphorylating enzymes are known collectively as *protein kinases*. The cAMP-activated protein kinase is abbreviated as PKA, to distinguish if from a calcium-activated enzyme, PKC. And many of the proteins they phosphorylate are themselves enzymes, whose activity can be regulated by phosphorylating or dephosphorylating (removing the phosphate) them. We will discuss the details of this reaction in the specific context of glucagon and adrenalin shortly; for the moment it is enough to note that all known effects of cAMP result from the activation of protein kinases.

PKA is composed of two kinds of subunits, a regulatory (R) subunit which can bind cAMP, and a catalyic (C) subunit. In the absence of cAMP, the regulatory and catalytic subunits form an R_2C_2 complex that is catalytically inactive. The binding of cAMP to each of the regulatory subunits leads to the dissociation of the R_2C_2 complex into an R_2 subunit and two C subunits. These free catalytic subunits are then enzymically active. Thus the binding of cAMP to the regulatory subunit relieves its inhibition of the catalytic subunit. Cyclic AMP acts as an allosteric effector (Chapter 5).

PKA and PKC are fairly universal classes of enzyme. The interaction between protein kinase, phosphorylation processes and Ca^{2+} turns out to represent a very general regulatory mechanism, switching on and off a variety of intracellular reaction sequences. One of the most interesting of these is raised by the fact that changing intracellular Ca^{2+} and cAMP concentrations can also provide nuclear signals to switch on and off particular genes by binding to DNA at a site called the *cyclic AMP response element*, itself requiring a specific binding protein, and hence functioning as a transcription factor in the way described earlier in this chapter.

Thus cAMP seems to serve as an almost universal signalling molecule, carrying diverse messages within different cells in different organisms. The fact that cAMP evolved to serve as a second messenger seems to have been favoured by four features of the molecule. Firstly, cAMP is derived from ATP, itself a ubiquitous molecule, in a simple reaction driven by the subsequent hydrolysis of pyrophosphate. Secondly, cAMP is not itself involved in any major metabolic pathway and is used only as an integrator of metabolism, not as a precursor or intermediate in energy production. Thus its concentration can be independently controlled. Further, it is stable unless hydrolysed by a specific phosphodiesterase. And finally, the complex molecular structure of cAMP enables it to bind specifically and tightly to receptor proteins, such as the regulatory subunit of protein kinase in muscle, and to elicit allosteric effects. More than a dozen different hormones use cAMP as a second messenger; as well as adrenalin and glucagon these include: calcitonin, chorionic gonadotrophin, corticotrophin, noradrenalin, parathyroid hormone, thyroid-stimulating hormone, and vasopressin.

It is important to appreciate that the hormonal signal is greatly amplified by the use of cAMP as a second messenger. The concentration of many hormones in the blood is of the order of $10^{-10}M$, that is, one ten-billionth of a mole per litre. In a stimulated target cell, the concentration of cAMP is very much higher, many cAMP molecules being synthesized by a single activated adenylyl cyclase molecule. The phosphorylation of many protein molecules by a protein kinase activated by cAMP increases the amplification, thus producing a cascade effect. The enzymatic cascade in the control of glycogen metabolism (page 287) illustrates how small stimuli can trigger major changes in cells.

When cAMP is involved as a second messenger, the effect of the hormone tends to be a large and long-lasting response, but because of the several steps involved a delay often occurs between receptor binding and the ultimate cellular response. Some hormones, however, can evoke immediate though transient effects, and in this case it seems likely that it is not cAMP but Ca^{2+} which itself serves as the second messenger. Hormone–receptor interaction can lead to a change in membrane fluidity which in turn leads to the opening of a pore through the membrane, allowing Ca^{2+} ions to flood into the cell. Continuing changes in fluidity will result in the almost immediate closing of this pore, which is known for this reason as a calcium gate. Some of the many regulatory effects of calcium have already been referred to, but amongst others not yet mentioned are its effect in causing muscle contraction, a process called stimulus-contraction-coupling (see the next chapter). Calcium has also been implicated in the secretion of packaged cell products such as hormones and digestive enzymes. It is also thought to be involved in the process of cell division, an effect known as stimulus-division-coupling.

It might be thought that with cAMP and Ca^{2+}, and their interactions, there were already quite enough second messengers to worry about, but if there is any general message to be drawn from modern biochemistry, it is that nearly every process studied turns out to be a bit like a Russian doll, opening up to reveal another and yet another inside it. Thus it turns out that there is at least one other major class of second messenger systems, based this time on a lipid component of the cell membrane, a multiply-phosphorylated inositol compound (page 79), phosphatidyl inositol 4, 5 biphosphate, or *PIP*$_2$. First messengers using this system bind to the cell membrane and activate a Ca^{2+}-dependent phospholipase enzyme, which splits PIP$_2$ to release two fragments, *inositol triphosphate* (*IP3*) and *diacylglycerol* (*DAG*). Both these last two can serve as second messengers; IP3 moves to the cytosol, where it serves to mobilize calcium, interacting with calmodulin in processes important for nerve cell activity and muscular contraction. DAG, which remains membrane-bound, activates membrane protein kinases which themselves affect the entry of ions into the cell. The basic reaction generating IP3 and DAG is as follows:

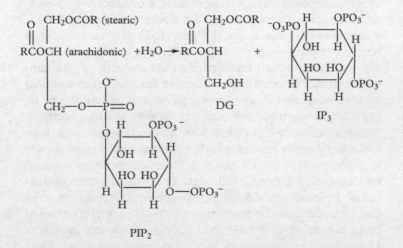

PIP$_2$

There is, as we have now learned to expect, a complex cycle of reactions, still only partially unravelled, which then resynthesizes PIP$_2$ – the inositol-lipid cycle.

The whole picture of hormone action is further complicated by the fact that many cells have receptors on their surface for different hormones, linked to different second messenger systems. In some cases the different systems have been shown to act antagonistically towards each other, providing a complex control mechanism for modifying the cellular response to hormones. For instance, not only does Ca^{2+} activate the phosphodiesterase responsible for the hydrolysis of cAMP, but conversely cAMP is thought to speed up the membrane-pumping mechanism that reduces intracellular Ca^{2+} levels.

Adrenalin, insulin and the control of carbohydrate metabolism

A variety of hormones affect the body's carbohydrate metabolism; some increase carbohydrate reserves by promoting the synthesis of glycogen, others accelerate the rate of glycogen breakdown and glucose oxidation. Between them, they add formidable weapons to the cell's already powerful armoury of glucose control systems.

The hormones mainly responsible are produced by two endocrine glands, the pancreas and the adrenals. The pancreas is a large gland situated in the duodenal loop of the intestine and ninety-five per cent of its tissue is devoted to the secretion of digestive juices which are carried to the intestine by the pancreatic duct. The other five per cent of the tissue (the so-called islets of Langerhans) contain several distinct cell types. The α-cells are responsible for the production of the hormone *glucagon* whilst the *β*-cells make *insulin*. Insulin is a low-molecular-weight protein which we have already met as the first ever to have its full amino acid sequence determined. In patients suffering from insulin deficiency (usually through failure of the insulin-producing *β*-cells), glucose circulates in large quantities through the bloodstream but is not taken up by muscle or liver either for oxidation or for conversion into glycogen. Ultimately, it is disposed of by excretion into the urine. Thus the patient suffers from a diminished carbohydrate metabolism despite the existence of a plentiful supply of glucose, and the body has to resort instead to fats and proteins to make up the energy-deficiency (resulting in the condition known as 'fatty liver' – see page 182) The insulin-deficiency illness is known as diabetes (or, more precisely, as there are several forms of this condition, diabetes mellitus).

Insulin, which interacts with receptors on the cell surface, promotes anabolic processes and inhibits catabolic ones in muscle, liver and adipose tissue. Specifically, it increases the rate of synthesis of glycogen, fatty acids and proteins, and stimulates glycolysis. It promotes the entry of glucose, some other sugars and amino acids into muscle and fat cells, hence lowering the level of glucose in the blood *(hypoglycaemic effect)*. Insulin inhibits catabolic processes such as the breakdown of glycogen and fat, and decreases gluconeogenesis by reducing the activity of enzymes such as pyruvic decarboxylase and fructose-1, 6-diphosphatase. In many ways its effects can be regarded as antagonistic to those elicited by adrenalin and glucagon, insulin signalling an abundance of glucose, adrenalin and glucagon its scarcity.

The work of Banting and Best in demonstrating that the injection of insulin into diabetic patients results in an immediate uptake of glucose into the tissues and a normalization of carbohydrate metabolism was mentioned in Chapter 6. Excessive insulin, on the other hand, results in so great a flow of glucose into liver and muscle that other organs, and in particular the brain, actually become glucose starved. Such starvation rapidly produces coma and even death, if not immediately alleviated by massive glucose injections. Insulin production must therefore be carefully balanced to avoid either too little or too great an uptake.

Adrenalin is a complex amine molecule produced by the central core region (medulla) of the adrenal gland. The hormone is secreted as part of the body's response to a new or potentially dangerous situation – adrenalin has been called the 'fight or flight' hormone.* Under its influence the heart beats faster, more blood flows to the muscles, several of which begin

---

*Actually 'fight, flight or frolic' but the puritanical English tradition usually omits the last. In the USA adrenalin is called epinephrine.

to contract, and the condition which we recognize in ourselves when we prepare to meet an emergency – a contracted stomach, muscular tension, and general 'nerves' – is produced.

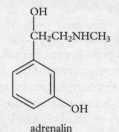

OH

CH$_2$CH$_2$NHCH$_3$

OH

adrenalin

A variety of these effects of adrenalin are mediated through the nervous system in ways that take us outside the scope of this book to discuss. So far as glucose metabolism is concerned, its role, as we have already seen, is to stimulate glycogen breakdown in muscle and liver. In this respect its function is similar to glucagon although they are structurally quite different, glucagon being, like insulin, a peptide. Insulin, as we have already said, has an opposing role to the other two hormones, in stimulating glycogen synthesis. We have already described how the increased intracellular level of cAMP resulting from adrenalin or glucagon triggers a series of reactions that activate glycogen phosphorylase, the initial step in the breakdown of glycogen to glucose (page 181) and inhibit glycogen synthetase. Both the phosphorylase and the synthetase exist in two forms, an active and an inactive state. In each case the hormone exerts its effect by phosphorylating the enzyme, the difference being that phosphorylation results in the production of the active form of glycogen phosphorylase but the inactive form of glycogen synthetase. It was this reaction which Sutherland and his colleagues were studying when they were led to the discovery of cAMP.

In glycogen breakdown there are three enzymically catalysed control stages, whereas in glycogen synthesis there are two control stages. Each step magnifies or *amplifies* the effect of the preceding one and the whole reaction is therefore known as a cascade system. It is illustrated in Figure 43. If the enzymes were directly regulated by the binding of adrenalin, more than a thousand times as much would be required to produce the glycogen breakdown as is needed in this cascade system.

If this seems complex, it is still only a simplification of what appears to happen *in vivo*. The enzyme responsible for phosphorylating the glycogen phosphorylase, *phosphorylase kinase*, is itself capable of existing in an active and an inactive form, and there should now be no prizes for guessing that the difference once again lies in the state of phosphorylation of the phosphorylase kinase. So there is yet another enzyme which activates phosphorylase kinase by phosphorylating it, and one which deactivates it by dephosphorylating it!

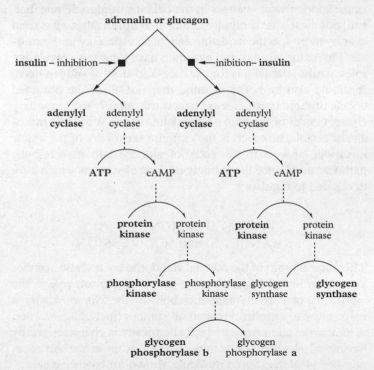

Figure 43. Hormonal control of glycogen metabolism–the cascade system

The interplay of these enzymes and their hormonal regulators by mutual phosphorylation and dephosphorylation constitutes one of the most complex and best studied control mechanisms in mammalian systems. The whole process is summarized in Figure 43 (where the bold type indicates the phosphorylated, active form of the enzyme) and which also shows how insulin acts by inhibiting both pathways simultaneously.

It is worth emphasizing that one of the important differences between the two hormones, and also between adrenalin and the thyroid gland hormone thyroxine, which also stimulates metabolism, is in part the immediacy of the action of adrenalin; the adrenal gland releases it in brief bursts which cause sudden swift changes in the balance of glucose metabolism, but these effects rapidly diminish as adrenalin itself is soon destroyed by specific oxidizing enzymes, especially in the muscle. Thyroxine, insulin, and glucagon have long-term regulating roles, whilst adrenalin is for the here-and-now of sudden need. It should also be borne in mind that not all of the observed effects of these hormones are necessarily simply the secondary consequences of the glucose-regulating steps. What this mechanism reveals, however, is the extreme sensitivity of the whole metabolic process to the entry of glucose into its metabolic pathway, and hence the subtlety of control systems which have developed to regulate it.

STEROID HORMONES AND ENHANCED GENE EXPRESSION

The outer margin of the adrenal gland, the cortex, also secretes another group of hormones that play a significant role in the regulation of carbohydrate metabolism, the *glucocorticoids*, a name given to a group of related substances (including *cortisone, corticosterone* and *cortisol*) which chemically is characterized by having an alcoholic or ketonic oxygen atom at carbon atom number 11 of the steroid molecule, shown for cortisone here:

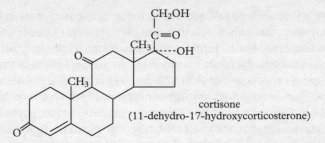

cortisone
(11-dehydro-17-hydroxycorticosterone)

One of the main effects of glucocorticoid hormones is to promote gluconeogenesis and the formation of glycogen in the liver. In addition, they increase the breakdown of proteins and amino acids. In general, they seem to operate at least partly so as to reverse or counterbalance the effects of insulin, so that, in a normal healthy animal or human, carbohydrate metabolism is controlled partly as the result of a tug-of-war between insulin on the one hand and the glucocorticoids on the other.

The fact that the glucocorticoids act so as to increase both glycogen synthesis and protein breakdown would suggest that they worked by speeding the conversion of amino acids to carbohydrate, and this has indeed been demonstrated by studying the fate of radioactive amino acids under the influence of these hormones. Glucocorticoids were found to have a direct effect in increasing several-fold the amounts of certain enzymes of amino acid breakdown, in particular transaminase and amino acid oxidase in the liver. The increased activity of these enzymes speeds up the rates at which amino acids are converted into urea and tricarboxylic acid-cycle intermediates such as α-oxoglutaric acid. The urea is excreted and the α-oxoglutaric acid can be either oxidized or used to resynthesize glucose and glycogen by pathways we have already described.

Thus these hormones are particularly important during prolonged starvation when glucose reserves are used up and the cell has to turn at last to the proteins. How do they achieve these effects? Because steroids are hydrophobic molecules, they can penetrate the membrane. Thus, although some steroid actions are also mediated by membrane-bound receptors, they have a further mode of action, via intracellular receptors in the cytoplasm, through which they can directly affect transcription of a large number of target genes. For instance cortisol, which is released into the bloodstream from the adrenal gland at times of stress or intense physical activity, will reach many different cells. Some contain *glucocorticoid receptors*, to which the cortisol will bind and the hormone-receptor complex is transported to the nucleus and binds to a DNA response element. In the liver, the result is the promotion of a range of enzymes necessary to increase energy production, including for instance those necessary for the conversion of amino acids to glucose via the enzyme transaminase. In the case of another steroid hormone, progesterone, binding to the DNA leads to the dissociation of the receptor-hormone complex into two subunits, one of which remains bound to nuclear protein, whilst the other binds to the DNA. Steroid receptors are widely distributed in the body; specific brain regions for instance are rich in glucocorticoid receptors whilst other regions contain receptors for the sex hormones oestrogen and progesterone. The role of the brain's glucocorticoid receptors seems to be to signal that the animal is in a stressful environment and requires the full attention of all relevant parts of the nervous system. Thus research I myself have done with colleagues in the Brain and Behaviour Research Group at the Open University has shown that one effect of corticosterone in the brain is to increase the synthesis of a class of membrane glycoproteins (cell adhesion molecules, page 93) which alter the strength of connections between nerve cells. However, as in all such matters, the exact amount of the hormone is important. Too much and nerve cells may die, an effect that seems to speed up ageing.

Regulation of other hormones

We have described a variety of regulatory influences exerted by a number of different hormones, which between them are capable of modifying a wide range of cellular activities. Several of these hormones, indeed, have mutually contradictory effects, and the final state of the cell must depend on the interplay between them and the relative amounts of each present. Either overproduction or underproduction of a hormone may lead to undesirable consequences: deformity, disease, coma, or death. Clearly, then, there must be some means whereby the healthy body can regulate the regulators, ensuring that the quantities of each hormone present at any one time are appropriate to the metabolic demands of the whole animal.

This regulation is achieved by a device of breath-taking simplicity. As each hormone is itself the product of biosynthetic pathways in particular endocrine organs, the amount present in the bloodstream may be controlled by an alteration of the rate at which it is being synthesized. And this rate is controlled for nearly all the major hormones (the main exception is insulin) by *other* hormones, produced in the pituitary. Thus the pituitary makes a set of peptide hormones whose target organs are the other endocrine glands of the body. These secondary, higher level regulators are called *trophic hormones;* they include thyrotrophic, adrenocorticotrophic, and gonadotrophic hormones, amongst others, and each has the property of stimulating the synthesis and release of the primary hormones from their respective glands.

But there is a possible dangerous trap here. What regulates the production of the pituitary hormones? Are we in danger of demanding an infinite regress of hormones, as once we found ourselves involved with an infinite regress of enzymes? There are two parts to the answer to this question, which between them reveal the full beauty of the regulatory mechanisms which the body has evolved. The first, and most straightforward, is that the rate of production of the trophic hormones themselves is dependent on the concentration of the hormones of their target glands.

For example, during the female reproductive cycle in mammals, the development and mobilization of the egg cell (oocyte) is controlled by two hormones produced by a particular region of the pituitary, Follicle Stimulating Hormone (FSH) and Luteinizing Hormone (LH). LH also stimulates the follicle cells around the developing ooctye to produce oestrogens which help prepare the placenta for pregnancy and which also inhibit FSH. If conception occurs, then the placenta takes over oestrogen production, ensuring continued inhibition of FSH and therefore making sure further ovulation does not occur. If there is no pregnancy the inner lining of the womb is shed during menstruation. The follicle cells regress at the same time so that oestrogen levels decrease, and FSH levels rise, initiating another cycle of ovulation. FSH is not only subject to inhibition by oestrogen but also comes under the control of an FSH-releasing hormone (FSH-RH) produced in another part of the pituitary. FSH-RH is also subject to feedback inhibition by oestrogen and FSH is in fact controlled by the combined effect of the two hormones.

Similarly the presence of thyroxine inhibits thyrotrophic hormone production and thyrotrophic hormone stimulates

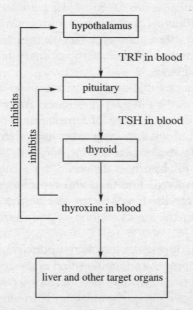

Figure 44. Hormonal feedback and control of thyroxine production

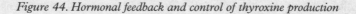

thyroxine production. This can be shown diagrammatically as a feedback loop (Figure 44). Hormone production thus forms a self-regulating system efficient and simple enough to gladden an information scientist's heart. However, Figure 44 also shows that the system is not entirely autonomous: the pituitary lies at the base of the brain, and it is from here that the second part of the answer to the question of what regulates the production of pituitary hormones comes. There is a direct blood supply linking a small region of the brain, the *hypothalamus*, with the pituitary. Down this path other peptide hormones, known generally as *releasing factors*, flow, synthesized in the hypothalamus and secreted as a result of the signals arriving at the hypothalamic nerve cells. It is these releasing factors which serve to modulate the release of the pituitary hormones.

GROWTH FACTORS AND OTHER MESSENGERS

The hormones form one group of what more generally are known as messenger molecules. As will become clear in the next chapter, communication between nerve and nerve, and nerve and muscle, is by way of *transmitter* molecules which diffuse across the junctions between the cells. Many of these transmitters are chemically similar to peptide hormones, others are amino acids. At least one, however, originally derived from endothelial cells, turns out to be in a class of its own, being a simple gas, the substance *nitric oxide, NO*. NO is synthesized from the amino acid arginine, by way of the enzyme *nitric oxide synthase*, and released across the cell membrane, where it quickly diffuses to other cells in the vicinity. Although it has a very short half life, being soon destroyed, it can cross the target cell membrane, interacting with G proteins and affecting calcium fluxes, thus triggering second messenger cascades. NO has important signalling functions in a variety of tissues. Until the late 1990s its role in the brain was thought to be amongst the most interesting, but then the drug company Pfizer launched Viagra, whose biochemical effects are to reduce the rate at which NO is destroyed and whose physiological consequence is to enhance penile erection and hence the quality of penetrative sex, ensuring it more publicity than any previous drug, even Prozac.

The final class of substances which can interact with the cell by way of membrane receptors are known collectively as *growth factors*. Their discovery dates back to the early 1950s, when the Italian biochemist Rita Levi-Montalcini was working in the US with tissue cultures of peripheral nerve cells. She found that whilst the cultures would maintain themselves alive in a medium which supplied them with the right combination of ions, glucose and so forth, they did not show much growth. However, when they were also supplied with an extract derived from mouse tissue, and especially its salivary glands, they would engage in a dramatic growth spurt, putting out nerve fibres which would extend energetically towards the highest concentration of the added extract. The extract thus possessed a *trophic* effect, not merely stimulating growth but also affecting the direction of growth.

It took many years before this curious observation was better understood, and it was not until 1986 that Levi-Montalcini was belatedly awarded a Nobel Prize, one of only ten women scientists to achieve this distinction since the prizes were instituted at the beginning of the twentieth century. The mysterious substance in the salivary glands extract was finally purified and christened *nerve growth factor (NGF)*. It turned out to be a small protein consisting of two identical polypeptide chains each with a molecular weight of 13000 and an amino acid sequence rather similar to that of insulin. Prior to its release NGF is stored as a much larger prohormone in a variety of sites across the body. It is now known to be but one of a large number of such small protein factors, derived from a variety of different tissues, and generally known from the tissue from which they have been purified rather than that at which they act. They include, for instance, Brain-Derived Neurotrophic Factor and Platelet-Derived Growth Factor, amongst others. Like the hormones, growth factors exert their cellular effects by interacting with membrane-bound receptors which also contain an amino acid capable of being phosphorylated, in this case tyrosine, and each growth factor interacts with a specific *tyrosine kinase*.

Growth factors exert a range of stimulatory effects on cells, both *in vivo* and *in vitro*. They are probably most active during development, when cells have not only to multiply but also to migrate to appropriate places in the growing organism, and gradients of growth factor, secreted by target organs, can help direct this growth and enable the migrating cells to find their way through a maze of other rush-hour travellers. However, as the nature of being alive as a multicellular organism is the constant birth and death of cells, and their modulation by experience and environmental circumstance, it is becoming clear that such factors have important functions to fulfil even in maturity.

CELLS IN CONTEXT

This chapter has begun the task of moving beyond considering cells in isolation, to regarding them, in multicellular organisms, as parts of a larger whole. Cells are organized into tissue systems and organs, each performing a particular task for the general well-being of the body. Not only do the cells in any one organ have to function together, they must also be aware of, and responsive to, events happening in other parts of the body. This led us to a discussion of the hormones by which the body controls and integrates the metabolism of its individual cells. So we came to a view of cells as interacting units in a greater whole, the entire organism.

In studying the finer aspects of the regulation of the cell, we found it necessary to take into consideration its external environment, full of challenge for single-celled organisms like bacteria, but in multicellular creatures protected by the body's homeodynamic system which maintains the constancy of its internal composition and thus enables each cell to survive as part of an integrated whole. Departure by very much from the norms of this uniform environment spells speedy death for any body cell.

But the body also has an external environment, and this, far from being the warm dark womb in which it cradles its own cells, is a tough, highly changeable, and often dangerous jungle. It is a jungle of rapid alterations in temperature, humidity, and consistency, in which food, instead of being wafted effortlessly along a swift-flowing bloodstream, must be hunted down energetically and captured by skill. The body needs to maintain homeodynamics, but in order to do so it has to convert the non-homeodynamic systems of the world around it. It has to act on this world in order to mould it into favourable contours. If temperature changes, the organism must organize itself so as to counter these changes, by reducing its heat loss, increasing its heat production, warming its surroundings or moving to more sheltered ones. It must be constantly in search of food, of air, of water, in order to meet the insatiable demands of its own interior.

And in this search it must inevitably conflict with other life-forms which are driven by the same internal urges to hunt for the same ends (for instance, organisms which are bent on devouring it). This need to act upon the world around it demands a biochemical and physiological integration of the body at the highest level, and it is to that which we must now turn if we are to achieve the biochemist's goal of describing the *total* behaviour of living organisms in physical and chemical terms.

CHAPTER 12

The Cell in Action

For an animal to act on the world around it, three abilities are required. In the first place, it must have the means of receiving stimuli from outside which will enable it to assess its position in space and its relationship to other objects; it must be able to see, hear, scent, and feel food and danger; i.e. it must have sense receptors, eyes, ears, nose. Second, it must be able to recognize, coordinate, and respond in an appropriate way to incoming sense stimuli. This task, in vertebrates, is carried out by the brain and central nervous system. Third, it must have a means of achieving purposeful movement in order to obtain a given goal, whether the capture of food or the avoidance of harm; this movement is produced by muscles. We can sum up this three-cornered relationship in a simple diagram (Figure 45).

By now, we should have no difficulty in framing the biochemist's question: what special chemical characteristics can we observe in these organs to account for their physiological role? This question has indeed existed for as long as biochemistry itself, and is, at one stage or another, even if it must be confined to their off-duty moments, on the lips of every biochemist. Ultimately, we only recognize life by its power to act upon the inanimate world around it, and our biochemical analysis cannot stop short merely at showing how the body, left to itself at a suspended interval of time, is composed and functions as an internally self-regulating homeodynamic system; in the long run, we must also hunt for the biochemical mainsprings of the body's relationship with the external world.

A good deal is known of the actual workings, biochemical as well as physiological, of nerve, muscle, and brain – much more than we have space to consider here. We must reluctantly leave out, for example, any description of the biochemistry of the special receptor cells of sense organs, although a study of each would reveal an exquisite interplay of biochemical and

Figure 45. Muscle, nerves, and brain

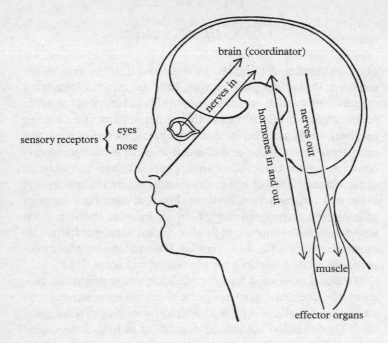

structural specialization which enables incoming information, in the form of pressure, vibration (sound), light, or indeed chemical composition, in the context of taste or smell, to be transduced into the universal language of the nerves, the passage of electrical impulses.

Here, we will begin first by considering the biochemistry of muscle, and then some aspects of the special biochemistry of nerve and brain.

MUSCLE

The typical skeletal (or voluntary) muscle, which is responsible for the movement of limbs and body, consists of a bundle of long fibres, each up to 0.01 centimetres in diameter and running the entire length of the muscle, joined at each end by tendons to the bone or organ they are responsible for moving. Movement takes place as a result of the contraction or relaxation of these fibres, which may shorten by as much as 60 per cent of their resting length, pulling the bone with them as they do so. Each fibre, in turn, can be seen under the microscope to consist of a set of long parallel fibrils, each about 0.0001 centimetres in diameter. These fibrils contain the contractile material of the muscle. Like other cells, muscle fibres have a nucleus, mitochondria, microsomes, and cell membrane, but when examining them under the microscope one is scarcely conscious of these; one sees the assembly of fibrils primarily as a long, thin, striped tube. These stripes are the most remarkable thing about skeletal muscle fibrils; they run across them at right angles at regular intervals, like the black and white markings on a zebra crossing. What is more, the stripes of adjacent fibrils all run in parallel, so that the stripes appear to continue right the way across the entire muscle fibre. It is this striped quality that particularly characterizes skeletal muscle, as opposed to certain types of smooth muscle which are not normally under voluntary control, but are responsible, for instance, for intestinal or stomach movements.

Viewed under a higher power microscope, the stripes may be resolved into alternate light and dark bands, forming the complex but regular patterns of Figure 46 (see also Plate 6). Microscopists refer to these bands as the I-(light) and A-(dark) bands, the H-zone (lighter area in the middle of a dark A-band) and Z-line. The one A-band, two half I-bands, and one H-zone between each pair of Z-lines represents a single unit, called a *sarcomere*. Each single unit is no more than 0.0002 centimetres in length, so to each fibril there are several thousand sarcomeres joined end to end. When the magnification is increased still further, using the electron microscope, the A- and I-bands are seen to be themselves composed of a set of filaments running parallel to the axis of the fibril. The A-band is dark because the filaments within it are thick; the I-band, on the other hand, consists of much finer filaments.

Figure 46. The muscle

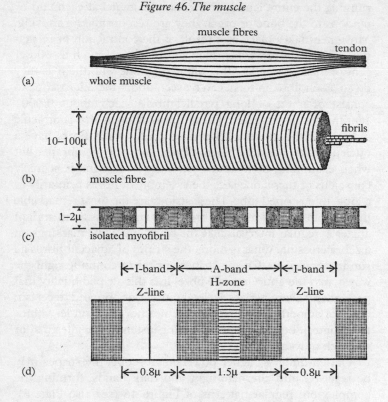

(a) whole muscle

(b) muscle fibre

(c) isolated myofibril

(d)

Muscle at different magnifications (1μ = 10⁻⁴ cm). Rabbit psoas muscle (after H.E. Huxley).

Careful examination shows that the slender filaments of the I-band run into the region of the A-band, filling the space between the thick A-band filaments, so as to form, in cross section, a regular pattern, each thin I-band filament being surrounded by three thick A-band filaments, whilst each A-filament has surrounding it six thin I-filaments. The I-filaments stop before they reach the centre of the A-band, so that the lighter-coloured H-zone of the A-band is in fact an area of the band which contains only thick A-filaments (Figure 47). Very

Figure 47. Muscle fibrils at high magnification

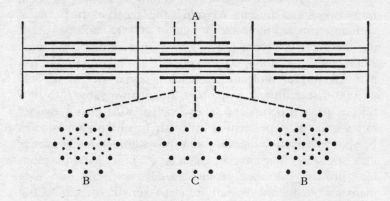

The arrangement of the filaments in the muscle fibrils. A: Longitudinal view. B and C: Cross sections showing thick and thin filaments (after H. E. Huxley).

close examination reveals that each I-filament, where it runs into the A-band, is connected with the A-filaments by fine, hair-like, cross-linkages (Plate 6.).

The filaments themselves do not change length when the muscle contracts; the sequence of events seen under the microscope is that the thin I-filaments slide in between the thick A-filaments until all the H-zone is filled up; when this happens the normal limit of contraction is reached. Thus the effect is that the I-bands seem to disappear whilst the A-bands remain unaltered, and the two Z-lines defining the limit of the sarcomeres are drawn in towards the central region.

The mechanism of muscular contraction, then, seems to demand the existence of a set of two types of interlocking filaments, contraction occurring when the filaments are slid together, relaxation as they move farther apart. In essence, muscular contraction is more like the shutting of a telescope than the tautening of an elastic band.

What can the biochemist add to this microscopist's and physiologist's view of muscle? It was early recognized that the contractile elements of the fibre were almost entirely composed of protein. There are at least four components, *myosin, actin, tropomyosin*, and *troponin*. Myosin is the largest of the proteins, with an estimated molecular weight of 460 000 and constituting the greatest proportion (about 54 per cent) of the total. Actin constitutes about 20 per cent of the total protein but it is a much smaller molecule with a molecular weight of only 41 000. X-ray diffraction patterns for myosin show it to be a fibrous protein composed of two polypeptide chains twisted into a helix for 90 per cent of its length. In addition myosin can be split by enzymic digestion into two subfractions known as light and heavy meromyosin. Actin on the other hand is a globular protein, but the globules readily assemble into long filaments. Actin and myosin are characteristic of muscle, but are also present in small quantities in many other cell types; for instance, nerves contain significant quantities of actin and it is clear that the proteins have a general function in cell movement and maintenance and alteration of shape, as well as their very special role in muscle.

The proteins can be separately extracted from muscle, and both types can be obtained as long fibrous filaments. In addition, myosin has the remarkable property of being capable of being extruded through a nozzle to form long artificial fibres (a process analogous to the formation of artificial fabrics such as nylon). When actin and myosin are extruded together, they form a complex fibre, actomyosin, and it was found by F. Straub and A. Szent-Györgyi in Hungary in 1942 that, if this actomyosin fibre is placed in a solution containing ATP and suitable salts, it will contract spontaneously, simultaneously dephosphorylating the ATP. Neither myosin nor actin by themselves, however, would behave like this.

It had already been known for some years that ATP and creatine phosphate were broken down during lengthy muscular contraction to release ADP and inorganic phosphate. Indeed it was these observations, made with muscle in the early 1930s, that first led to the recognition of the role of organic phosphates in biochemical reactions. In 1933, V. A. Engelhardt in Moscow was able to show that, in the presence of myosin, ATP was rapidly split to ADP and inorganic phosphate. Thus myosin seemed to function as an ATP-ase enzyme. It was clear that these observations must be related in some way to the explanation of muscular contraction. The situation was further complicated though by the observation that the injection of a tiny amount of Ca^{2+} into a single muscle *in vivo* will produce a local contraction although the purified actomyosin complex does not require Ca^{2+} for its contraction.

However, it required the coming of the electron microscope and the techniques of X-ray crystallography to be able to link decisively the biochemical and microscopical observations, and these later interpretations were to a large extent the work of H. E. Huxley and A. F. Huxley and their co-workers in England during the 1950s. The simplest interpretation is that the two major proteins, myosin and actin, correspond to the two different types of filament in the muscle fibril. The thick filaments of the A-band are composed of myosin, the thin I-band filaments are actin. The myosin molecules in the thick filaments form rods composed of two chains. Each chain has a thickened portion at one end giving the rod a double-headed appearance. These 'heads' come into contact with the actin molecule and are composed of heavy meromyosin which carries the ATP-ase activity. The direction of the meromyosin heads reverses in the centre of the thick filament so that each filament is bipolar with a central bare zone, visible under very high magnification, containing no meromyosin heads. As we have said, actin is a globular protein and the thin filaments are formed by two strings or chains of actin molecules wound loosely together like two strings of beads. Tropomyosin is a fibrous protein which seems to lie in the groove formed by the winding of the two actin chains. Troponin is arranged at regular intervals along the tropomyosin molecule.

This structural information enables a contractile mechanism to be proposed. The current view is that the signal for contraction in the type of muscle we have been discussing is the release of a chemical transmitter from nerve endings at the junction between nerve and muscle, which we will shortly consider in more detail. This leads to a depolarization of the muscle cell membrane, which in turn results in the release of a large number of Ca^{2+} ions into the muscle fibre – the amplification process. These Ca^{2+} ions act as the second messenger and interact with the muscle proteins to bring about contraction. Thus the initial stimulus has been coupled to the contractile process by the Ca^{2+}. A possible mechanism is that the Ca^{2+} combines with troponin and this causes the tropomyosin in the same region to shift slightly, exposing a previously hidden part of the actin molecule.

Both the actin and myosin molecules carry strongly negative charges on their surfaces – indeed it is this electrochemical repulsion that helps maintain the structural integrity of the whole muscle. The minimum potential, that is, the most positive region, exists at the midpoint between the two filaments and this region is occupied by the meromyosin projections containing the multivalent and electronegative ATP system. The effect of Ca^{2+} binding to the actin filament will be to make the surface of the filament temporarily electropositive over a very localized area. The ATP complex will be attracted towards it and will interact. The ATP is then broken down to ADP with the resultant release of energy. Cessation of the nervous impulse causes an immediate reuptake of the calcium ions from the fibrils, the meromyosin-ATP complex is released back into the interfilament space and the tropomyosin-troponin component shifts back into the actin groove, but in a slightly different position, relative to its original alignment with the myosin filament. Thus the next site of attachment of the meromyosin molecule will be to a troponin molecule farther along on the actin filament, and hence the actin and myosin move relative to one another. As this process is repeated during contraction then the muscle shortens not by any change in length of the filaments but by the filaments sliding past each other, gradually filling the H zone. The process is summarized in Figure 48.

Figure 48. Structure of myofilaments and stimulus contraction coupling

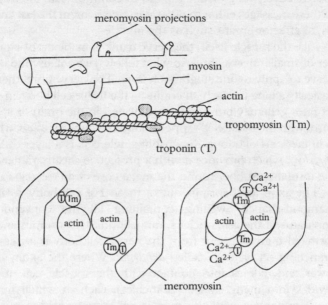

NERVE AND BRAIN

The great mass of cells of the central nervous system is concerned with the processing and coordination of the millions of incoming messages arriving every second from the internal and external sense organs and the peripheral nervous system, the control of the motor activities of the myriad of muscles, and the continued matching and comparing of the messages and responses which compare the present moment of existence of the organism with the accumulated wisdom of its past history, relating this to present needs and possible outcomes of actions. Of all that might be said of the biochemistry of nerve and brain at this point, we will confine ourselves here, by a process of tunnel vision, to just three issues: how the structures of the nerve cells are specialized to enable them to perform their functions; and what the biochemist can say about two fundamental properties of nerve cells, their capacity to conduct messages (impulses) over relatively long distances, and to communicate the import of these messages either to other nerve cells, or, in the last analysis, to effector organs such as the muscle.

Like the muscle itself, the nerve trunk is made up of a number of smaller nerve fibres; in what follows, we will use the word nerve to apply to one single such fibre. The nerve fibre, then, is basically a long tube which connects the nerve cell body, in vertebrates generally but not always located in the brain or spinal cord, with the muscle. Wrapped round this tube, or *axon*, which is in fact part of a long-drawn-out single cell, is a layer of lipid (*myelin*), which provides it with a protective sheath, rather like the insulating rubber round the metal core of an electric cable.

The axon itself springs out of the nerve cell body, a large, diamond-shaped swelling containing all the conventional apparatus of the cell; nucleus, mitochondria, ribosomes and so forth. Also growing out from the nerve cell body is a tree-like branching set of tubes called *dendrites*. Where the axon, at its lower end, comes into contact with the muscle cell, it too divides into many smaller branchlets, each of which makes separate contact with the muscle cell, thus touching quite a broad area of the muscle cell surface. Although under the microscope it is the nerve cell bodies (*neurons*) which catch much of the attention, in many nerve cells, so much do axons and dendrites branch, that they make up as much as 90 per cent of the cell's volume.

This region of contact between nerve and muscle is called the *neuromuscular junction*. The neurophysiologist's name for such a junction is a *synapse*. At the synapse, the tube which forms the axon swells as if being blown into a small balloon, to form a *synaptic bouton*. The membrane surface of the bouton spreads along the surface of the muscle fibril, to form a contact zone between pre- and post-synaptic cells. If, as frequently happens, the distance between the brain or spinal cord and the muscle is too long for one axon to stretch the whole way, a break is made en route between them. The nerve leading from the brain then forms a synapse with the dendrites or cell body

Figure 49

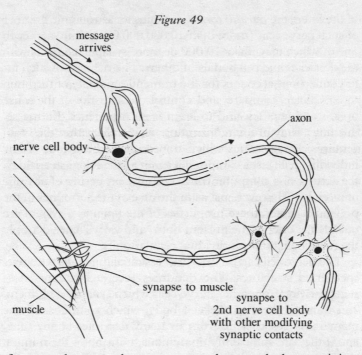

message arrives

nerve cell body

axon

synapse to muscle

muscle

synapse to
2nd nerve cell body
with other modifying
synaptic contacts

of a second nerve, whose axon can then reach the remaining distance to the muscle. Such a synapse between two nerve cells, as well as being a means of sending a signal over long distances, also, and more importantly, provides a point at which the signal message, descending from the brain, or arriving from a sense organ to the brain, can be modified as a result of information arriving from other sites. The message coming from the brain crosses the synapse between the first nerve and the second and continues down the second nerve to the neuromuscular synapse where it signals the muscle to contract. A similar system carries messages from sensory organs such as eyes and ears to the brain, and yet further synapses interconnect the large number of cells within the brain itself. Thus messages can be rapidly transmitted over considerable distances and to a large number of different nerve cells or muscles (Figure 49).

In mammals, the mass of neurons which wraps round the top

of the brain, the *cerebral cortex*, contains an astronomic number of such nerve cells (more than 10 000 000 000 in humans) each one of which may make 10 000 or more synaptic contacts with the dendrites and cell bodies of others. The richness of such interconnections accounts for the tremendous ability of the brain to coordinate, compare, and control the activities of the sense organs and muscles, and to learn from experience. En masse, this huge array of interconnecting nerve cells within the brain acquires characteristics which transcend the behaviour of any individual cell (just as, to use yet again a technological analogy, the activity of a computer transcends the properties of any one of its chips). However, just as in the case of the hormones in the previous chapter, these properties of the brain as a system are outside the brief of the present book, and we will consider only the functioning of its individual cellular units.

The nerve cell carries out three essential functions within its specialized structures. The dendrites receive incoming messages arriving from other nerve cells which synapse with them, and transmit them to the cell body, which computes all the many messages arriving from its many dendrites at any time, and if they are sufficiently unanimous, transmutes them into a single message which is passed swiftly down the axon to the synaptic bouton, where a signal is transmitted across the nerve cell membrane (here called the *synaptic membrane*) to the nerve or muscle cell at the other side of the synapse.

The simplest part of the process to understand is the way the messages pass down the nerve axon itself. Here our earlier description of the nerve fibre, surrounded by its lipid sheath like an insulated electric cable, comes strikingly into its own. In fact, messages pass down the nerve axon in the form of electric currents which begin where the axon springs out of the nerve cell body and travel down the fibre at a rate of about 20 metres a second.

Two sorts of job are done by more familiar electric currents. A steady current flowing to an electric light bulb provides a constant source of power for the bulb; if the flow of electricity alters, the bulb flickers. A door-bell, on the other hand, must

Figure 50. Voltage changes as impulse passes down nerve fibre

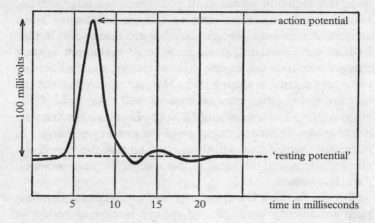

ring briefly and then stop. A short burst of electricity is wanted here, to act as a signal. The currents flowing in the nerve fibre are signals, not power supplies. Thus if we record the passage of electricity down a nerve we do not find a steady reading of, say, four volts, such as one might get from a torch battery; instead, at any one point along the nerve, the voltage suddenly fluctuates, rises to a peak, and declines again as the electric signal arrives, passes the recording point, and is gone. A wave of electricity has passed down the fibre. If we draw a curve plotting the rise and fall of voltage at a particular point on a nerve fibre during the passage of an impulse down it, we get the picture shown in Figure 50.

What is the mechanism of this remarkable phenomenon? In the sort of electricity produced by a generator or battery, the current running through the cables is in fact carried by a flow of electrons moving down the wire from the negative to the positive terminal. In animal-generated electricity, the current is still carried by a flow of charged particles, but, instead of the negatively-charged electrons, the carriers are positively-charged *ions*. That this flow is possible depends on a fact we have already commented on on more than one occasion – that the ionic composition of the inside of a cell is quite different from that of the outside (see page 93, for instance). Inside, there is a high concentration of potassium, but little sodium. Outside, there is a great deal of sodium but only a small amount of potassium. Nerves resemble most other cells in this. Sodium and potassium, as we know, exist in solution as their charged ions each carries one positive charge, and we have become accustomed to writing them Na^+, K^+. Of course, the cell contains many other charged ions as well: Mg^{2+} and Ca^{2+} amongst the positive ions, and Cl^- and phosphate amongst the negative ions. In general, there are about as many negatively as positively charged ions within the cell, so that the overall net charge is zero, and the same is also true of the fluid in which the cell is bathed.

But the fact that the concentrations of sodium and potassium are different either side of the cell membrane results in certain peculiarities despite this overall electrical neutrality. For in a non-living system, we would expect that if we separated a solution of potassium chloride from one of sodium chloride by a thin, permeable membrane – say a cellophane sheet – potassium ions would pass through the barrier in one direction, and sodium in the other, until the concentration of both ions was equal on either side of the membrane. That this does not happen in living cells is because the cell membrane behaves, as we discussed in relation to the general properties of the cell membrane in Chapter 4, as if it were impermeable to sodium. It permits potassium to enter the cell but not to leave, but does not let sodium enter at all. This is not entirely a passive refusal of admittance, for if sodium is injected into the cell, it is quickly pumped out again through the cell membrane *against* the sodium concentration gradient. The membrane is thus a dynamic system capable of selectively distinguishing between sodium and potassium ions.

What is the effect of these differences in concentration? If potassium is not to flow out of the cell down a concentration gradient, it must be in response to some other force acting inwards. This force is provided by the positively-charged sodium ions, which line up on the outside of the membrane so as to

Figure 51. Nervous transmission

(a) At rest, charged ions at either side of nerve membrane provide a potential difference.

out

| + | + | + | + | + | + | + | + | + | + | + | + |

| − | − | − | − | − | − | − | − | − | − | − | − |

in

| + | + | + | + | + | + | + | + | + | + | + | + |

| − | − | − | − | − | − | − | − | − | − | − | − |

out

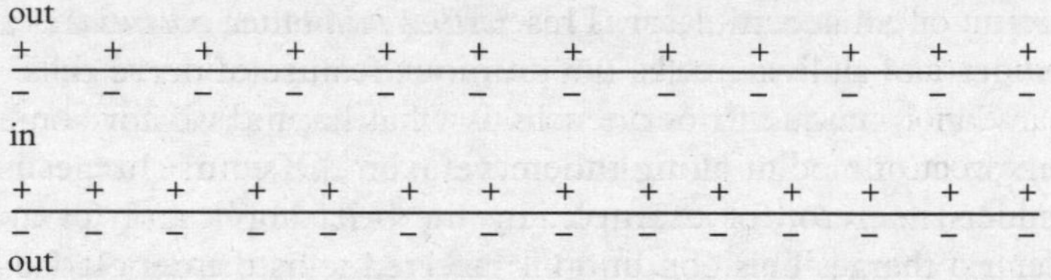

(b) A local depolarization at A results in current flowing, and the message passes down the nerve fibre.

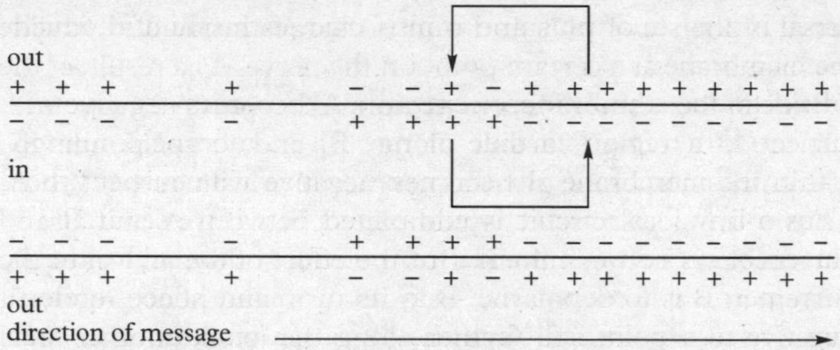

out

| + | + | + | | + | | − | − | + | − | + | + | + | + | + | + | + | + |

| − | − | − | | − | | + | + | + | − | − | − | − | − | − | − | − | − |

in

| − | − | − | | − | | + | + | + | + | − | − | − | − | − | − | − | − |

| + | + | + | | + | | − | − | + | − | + | + | + | + | + | + | + | + |

out

direction of message ⟶

provide a barrier of positive charge which repels the positively-charged potassium ions. At the same time the sodium ions tend to attract negatively-charged ions such as phosphate or chloride across the cell membrane. The net effect is that we can draw the axon as shown in Figure 51(a) with a line of positive charges down its outside surface and of negative charges on its inside. A potential difference thus exists between the two sides of the membrane – if they were shorted by a wire running across the membrane they would register a voltage. This can, in fact, be done, if a tiny glass microelectrode is inserted into the axon, and connected through a voltage recorder to the outside surface of the nerve. It is also possible to calculate a theoretical voltage which should arise in a system in which a membrane permeable to potassium but not sodium exists – a comparatively simple piece of mathematics. Fortunately, the theoretical and experimental results agree; the voltage recorder indicates that the inside of the cell maintains a potential anywhere between 65 to 95 millivolts negative to the external surface. The existence of such a voltage is identical in principle to the voltage produced by the dry chemical battery of a torch or the wet battery of an accumulator. This *resting membrane potential* is a property of all living cells, not a unique feature of nerve cells.

What *is* unique to nerve cells is what happens if for some reason at one point along the nerve axon the set of charges is reduced to zero, for example, by the brief application of an electric charge. This condition is referred to as the depolarization of the nerve, and its results are indicated in the diagram Figure 51(b). The results of depolarization are to cause a reversal of the set of plus and minus charges inside and outside the membrane at a certain point on the nerve. As a result, at the outside of the membrane, one region, A, becomes negative with respect to a region farther along, B, and, correspondingly, within the membrane B becomes negative with respect to A. Thus a tiny local circuit is completed between A and B and current flows between them. But the effect of the arrival of the current at B is to depolarize B in its turn, and hence render it negative to a point still farther along the axon; another local current then begins to flow and the depolarization is repeated. Thus a depolarization at one end of the nerve axon results in the establishment of a series of local circuits, which, moving like a wave over the surface of the axon, will carry the depolarization along its length. Provided the initial stimulus is large enough to set the circuits going, it is carried rapidly down the nerve fibre. This is called the *action potential*. This method of propagation of nerve impulses had been recognized ever since it became possible, early in the twentieth century, to record electrically from nerves, but the understanding of the physics and mathematics of the process is mainly due to A. L. Hodgkin, R. D. Keynes, and their collaborators in Cambridge in the 1950s, and is regarded as a universal mechanism, as fundamental for nervous transmission as, say, glucose oxidation is for ATP generation within the cell.

It was Hodgkin, too, who provided a satisfactory explanation of the phenomenon in biochemical terms. We have seen that the resting potential of the nerve is maintained by the difference in concentration and permeability of sodium and potassium. Suppose, Hodgkin argued, that the effect of depolarization is to cause a change in the permeability of the membrane so that it becomes for a brief while very permeable to sodium, what will happen? Sodium will flow rapidly into the nerve axon down its concentration gradient, entering faster than potassium can leave. As this happens the membrane potential will drop still further, as positive charges are transferred from outside to within. As the membrane potential drops further, it will become still more permeable to sodium, and the result will be a positive feedback system in which the entry of sodium is self-stimulating (Figure 52).

Figure 52

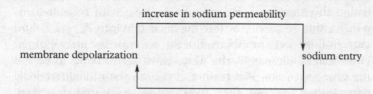

increase in sodium permeability

membrane depolarization   sodium entry

This inrush of sodium continues until the negative membrane potential of up to 95 millivolts is reduced to zero and then converted to one of some 20 to 30 millivolts positive. This represents the 'spike' of the wave of Figure 50. Sodium entry now ceases, as it is having to enter uphill against a potential gradient. At the same time, according to the Hodgkin model, the potassium permeability of the axon increases, and potassium leaks out down *its* concentration gradient until the membrane potential falls to its previous negative value and the electrical wave has passed on down the axon. In fact, potassium exit goes on slightly longer than is strictly necessary, and the potential slightly overshoots its original value, resulting in the second 'trough' of Figure 50. During this period, when the potential is some millivolts below the normal resting level, the nerve becomes incapable of carrying a second message. Within a few thousandths of a second, however, the old value has been restored once more and the nerve is again excitable and ready for action.

Hodgkin's theory has been amply borne out by experiments. Studies using radioactive sodium and potassium have demonstrated the changes in entry and exit of these ions across the membrane during nervous stimulation, whilst it has long been known that, just as the theory would predict, nervous conduction becomes impossible if all the external sodium is removed.

The changes in membrane potential are achieved at the expense of the sodium and potassium levels which the cell membrane normally maintains. The changes we have described as occurring at the membrane result only in rapid local changes of concentration, so that they hardly affect the overall differences in sodium and potassium levels inside and outside the cell. Despite the local changes, these concentration differences remain so great that, in a large nerve, upwards of a million impulses could be carried before the decline in internal potassium and rise in sodium became great enough to inactivate it. Nonetheless, ultimately, the nerve must set to work to rectify the changes in concentration and restore the old differentials between sodium and potassium. It may be noted that, until now, the processes we have described have not demanded the output of energy by the nerve; the changes in membrane permeability instead utilized the potential energy latent in the differences in sodium and potassium concentration across the membrane. The work of the nerve consists in establishing this potential energy gradient once more.

Again, the critical experiments which demonstrated the mechanisms of this process were performed by Hodgkin, by Keynes, and by A. F. Huxley. They used in their experiments the biggest nerve they could find – the giant axon of the squid, which is so wide, being up to a millimetre in diameter, that it is relatively easy to inject test substances down inside it and to study their effects. They first showed that nervous conduction is dependent on energy metabolism. Nerves poisoned with cyanide, which prevents oxidative phosphorylation, rapidly lose their ability to conduct impulses. But if ATP is injected into the poisoned nerve, it becomes active once more, and can go on conducting until all the ATP is broken down. Thus, just as in muscle, the source on which the cell draws for its ability to act is ATP.

Later work has been able to show that the ATP is in fact utilized in order to maintain the sodium and potassium levels of the nerve. ATP is converted to ADP during the extrusion of sodium and the entry of potassium. The entry and exit of these ions are found to be linked processes, and, in general, one molecule of ATP is broken down for about every three ions of sodium and potassium transported across the membrane. The mechanism is comparable to that described in the previous chapter for membrane-related processes, and involves phosphorylation of membrane proteins so as to induce conformational changes in carriers, or open pores through which ions can pass.

Figure 53. Transmission at the synapse

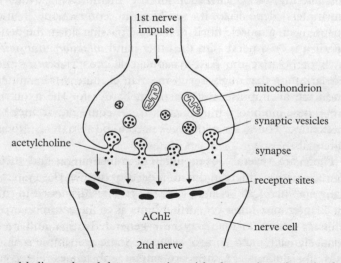

Acetylcholine, released from synaptic vesicles by an incoming stimulus, diffuses across the synapse to receptor sites in second nerve (or muscle) where it is destroyed by acetylcholinesterase, depolarizing the second nerve in process.

Transmission of impulses at synapses

The nerve impulse flows down the axon in a wave of depolarization of the membrane. But at the synapse a different problem arises. The membrane of the nerve cell or muscle fibril at the other side of the junction is not continuous with that of the incoming nerve. A means must be found of bridging the gap between the two cells. Essentially, this demands that the arrival of the depolarizing wave at the end of one nerve axon acts as the trigger or signal for the start of a similar wave in the nerve cell and axon of the second nerve or of the muscle. This triggering action depends on the existence of chemical *transmitters* present in the synaptic bouton. The arrival of the nerve impulse at the bouton stimulates the release of a chemical which can diffuse through the synaptic membrane and interact with the post-synaptic muscle fibril, dendrite or cell body (Figure 53). There, it stimulates the flow of ions across the membrane, thus altering the polarization of the membrane; *excitatory* transmitters depolarize it, resulting in contraction, if the synapse is on a muscle fibril, or a signal passing down the dendrite if it is on a nerve. On the other hand *inhibitory transmitters* hyperpolarize the post-synaptic cell, thus *preventing* the depolarization that might result from an excitatory transmitter producing an impulse and making it harder for the axon to 'fire'. These inhibitory transmitters are of equal importance to the excitatory ones, for nerve cells need to be able to say No as well as Yes.

There are several dozen different transmitters, and their chemical nature differs in different parts of the nervous system. Many are chemically related to the hormones discussed in the last chapter, providing yet further hints as to the common evolutionary processes that may have generated nerve and hormonal signalling and control systems. Some are simple amino acids, like glutamate. Others are amines, substances related to the hormone adrenalin. Still others are small peptides, some chemically related to vasopressin and other peptide hormones. One of the commonest is *acetylcholine*.

$$CH_3.C.OCH_2.CH_2N.CH_3.CH_3.CH_3$$

with the O double-bonded above the C.

While acetylcholine is the general excitatory transmitter in the periphery of the body, glutamate is probably the commonest in the brain itself, whilst amongst the important inhibitory transmitters is an amino acid found only in the brain, γ-amino-butyric acid or *GABA*. Another transmitter, the amine *serotonin* (5-hydoxy-tryptamine), is especially associated with neural pathways concerned with mood. A further, *dopamine*, is the transmitter whose synthesis fails in Parkinson's disease.

It seems to be the case that each pathway in the brain and nervous system, involving interaction and signalling between many nerve cells, utilizes a particular transmitter; thus pathways are chemically specified (like colour-coded wires in electrical equipment) as well as being linked by specific synapses. For example, in some of the brain pathways involved in emotional and attentional responses, the transmitters are amines, while some of the small peptide transmitters are involved in the transmission of pain sensations. There was much excitement following the discovery in the mid-1970s of a new class of these transmitters, *endorphins*, which were found to be chemically rather similar to the widely used pain-reliever morphine, thus providing a possible explanation for how such drugs might work.

Because many of the synapses between nerve and voluntary muscle are mediated by the excitatory transmitter acetyl-choline, this was one of the first to be studied in detail, and the general principles by which it works are the same for other transmitters too. Acetylcholine is manufactured by the acetyla-tion of the lipid choline by acetyl-CoA, and the responsible enzyme, *choline acetylase*, is present in the synaptic boutons. Both it, and the acetylcholine itself, are packed into small particles, visible in the electron microscope, called *synaptic vesicles* (Plate 7). The arrival of the nerve impulse causes many of the vesicles to move towards the synaptic membrane, fuse with it and empty their contents into the gap between the pre- and post-synaptic membranes, the *synaptic cleft*, through which they can diffuse to the second nerve cell or to the muscle. There are specific proteins located both in the vesicle membrane and the synaptic membrane itself which link transiently during this fusion process.

On the post-synaptic side, there are specific receptors located in the membrane on to which the transmitter binds, in a similar way to the type of hormone-receptor interaction described in the last chapter. The molecular structure and mode of action of many of these receptors, especially the acetylcholine and glutamate receptors, has been studied in considerable detail and it is now clear that there is not just one but many different receptor subtypes for each of the transmitters. In general, the binding of transmitter to receptor results in a change in the post-synaptic membrane structure. If the receptor is, like glutamate, an excitatory one, this may result in an influx of Ca^{2+} ions and an activation of second messenger cascades large enough for the post-synaptic membrane to become depolarized. If a sufficient number of synapses transmit excitatory messages to the post-synaptic nerve at around the same time, the result will be a general depolarization, and the second nerve will 'fire' or the muscle contract. On the other hand, the inhibitory transmitters, such as GABA, working in the same way, produce a hyperpolarization of the post-synaptic membrane (i.e. make it more, rather than less negative) – thus making it harder to fire.

Once it has exerted its effect, the transmitter must be destroyed to prevent its presence from interfering with further messages. In the case of acetylcholine, this is achieved by the enzyme *acetylcholinesterase*, which, present in the post-synaptic membranes, hydrolyses it again to choline and acetic acid. Other transmitters are reabsorbed after use, either into the pre-synaptic cells or into a further class of cell, the *glial* cells, which are present in nervous tissue in large numbers, surrounding most of the nerve cells and the synapses.

As many forms of psychic distress such as depression for instance are associated with problems of neurotransmitter metabolism, drugs designed to alleviate the symptoms often do so by interacting with transmitter synthesis, release, reception or reuptake. For example, a class of drugs known as selective serotonin reuptake inhibitors are widely prescribed for alleviating depression. One of the better known of these is sold as Prozac. Some of the mood-affecting *psychotropic agents*, like cannabis (marijuana) and LSD, may exert their effect by interacting in some subtle way with particular nervous system transmitters. And we have already referred to the euphoria and pain-reducing effects of morphine and morphine-like drugs, assumed to derive from their chemical similarity to the endorphins. The whole process of synaptic transmission is summarized diagrammatically in Figure 53.

The neurotransmitters like acetylcholine establish contact between two cells across the gap which separates them. It is not surprising to find that this is a very vulnerable link in the chain of signalling between nerve and muscle. Anything which interferes with the diffusion of the transmitter across this space will prevent the message being passed on, and many drugs which interfere either with transmitters, or their receptors, or the subsequent destruction of the transmitter by post-synaptic enzymes, or reuptake into the glia, thus block synaptic transmission and result in a sort of chemical paralysis. For acetylcholine, nicotine is one such drug; another is the poison which Central American Indians used to dip their arrows into and which has since become beloved of crime-writers, the mysterious curare.

In more recent times another class of substances which affect acetylcholinesterase has achieved more sinister significance, though. Originally a product of German research, a group of organophosphorus compounds which act in this way were developed as nerve gases during the 1939–45 war. An even more toxic variant (a milligram or so absorbed through the skin is lethal, making them amongst the most poisonous known substances) was developed in the mid-1950s at Britain's Chemical Warfare Station at Porton, in Wiltshire, the so-called

V-agents. (These are related to insecticides which are intended to poison insect but not mammalian acetylcholinesterase.) By poisoning acetylcholinesterase they prevent the removal of the acetylcholine at the post-synaptic sites; thus the message instructing muscles to contract is never countermanded and the muscles go into permanent spasm. Nerve gases like these were used first (and hopefully for the *only* time) against humans when the Iraqi regime of Saddam Hussein sprayed them on the Kurdish population of the village of Halabja in northern Iraq in 1987.

If the internal working of our cells depends ultimately on ATP, the ability to control and command our body to act is similarly dependent on the neurotransmitters.

CHAPTER 13

Defending the Organism

Up till now we have considered the animal cell as a homeodynamic system, with autonomous properties and behaviour, but also as part of a greater whole, the organ and the organism in which it has a specialized function. Cells grow, mature, maintain themselves and act on their environment; they make and secrete hormones, or contract, or transmit and process information. But in all this we have considered the environment of the cell as at best a beneficent or at least a neutral medium, a vehicle through which nutrients and oxygen are supplied and waste products removed. Yet of course such an account fails to describe the full complexity of the world in which cells and organisms find themselves. Organisms interact with and prey on one another. Each organism, going about its legitimate business of surviving and reproducing, finds itself sometimes in cooperation, sometimes in competition, with others. To survive in an environment which may be hostile as well as supportive, organisms, and their cells, need to be able to defend themselves against assault. For the whole organism, this may mean developing jaws or claws, shell or horn, protective coloration or speed of running, stings or bitter taste. These may protect against larger enemies, but not against smaller ones such as many bacteria and viruses which can enter the organism and subvert it from within. To defend against these sources of harm, or *pathogens*, whose entry can only be limited but never entirely prevented, an organism needs a defence system capable of protecting its host from infection by *recognizing*, *destroying* and *eliminating* invading pathogenic microorganisms. This is the function of the body's *immune system*, to the biology of which we now turn.

Folk knowledge that once having been mildly infected with a disease may confer some protection against reinfection, runs back as far as or beyond history; ancient civilizations like the

Chinese knew, centuries before Jenner, that if you had once had cowpox you would be immune to smallpox, but understanding of the molecular and cellular processes involved in this protection – and hence the science of *immunology* – is contemporaneous with modern biochemistry and genetics. Immunology is a complex science with a language and methodology which only partially overlaps with the biochemistry we have discussed up till now, and this chapter is limited to discussing those aspects of immunological mechanisms which flow from biochemistry and molecular biology.

Immune mechanisms vary enormously in complexity in different species. All begin, however, with the capacity of cells to recognize and react to macromolecules which are 'foreign' to the organism. And recognizing that some substance is foreign must mean that an individual's cells can recognize one another as being *non*-foreign but in some sense like themselves. At the heart of the immune response, therefore, lies a capacity of the immune system to distinguish 'self' from 'non-self'. These are of course somewhat mysterious categories to philosophers and moralists, but ones which in most circumstances animal cells have little difficulty in making judgements about. Invertebrates can recognize foreign transplants – a phenomenon known as *innate immunity*. Vertebrates, however, whilst retaining this power, have developed in addition a more sophisticated set of mechanisms which are known as *adaptive immunity*. The term innate implies that these mechanisms are present from birth and are rather automatic, non-specific and inflexible responses which do not alter on repeated exposure to a given infectious agent. By contrast, the adaptive immune response is very specific for a given pathogen and improves with each successive encounter with the same pathogen. In effect, the adaptive immune system is capable of 'remembering' the infectious agent and thus can prevent it from causing disease later. This makes it the basis of the broadly used methods of protective vaccination against infectious diseases.

INNATE IMMUNITY

Effector cells

Cells involved in mediating innate immunity are endowed with the capacity both to recognize and to respond to invaders. That is, to engulf bacteria and viruses and to activate enzyme systems and generate toxic molecules, such as reactive oxygen and nitrous oxide intermediates, peptides and enzymes that can kill invaders; to secrete *cytokines* which help produce inflammation; and to migrate, adhere to, and penetrate tissues. These effector mechanisms together contribute to both local and systemic effects that mediate host defence. Three kinds of white blood cell are responsible for destroying pathogens by *phagocytosis* – that is by swallowing or engulfing them. These cells, known collectively as *phagocytes*, are the *neutrophils*, *monocytes* and *tissue macrophages* and work in similar ways. Specialized proteins present in the blood first bind to a pathogen and 'label' it chemically as a target for phagocytosis. These chemical labels are known collectively as *opsonins*. The coating of pathogens with proteins that promote phagocytosis is called *opsonization*, and the process dramatically speeds the rate of elimination of pathogens from the bloodstream. Once the phagocyte has attached itself to a pathogen a system of contractile actomyosin filaments inside it is brought into play so as to wrap the phagocyte's membrane round the target pathogen. The membrane surrounding the pathogen then fuses with the membrane surrounding lysosomes inside the phagocyte. Thus the pathogen is pulled into the lysosome whose hydrolytic enzymes will, as will be recalled from Chapter 4, be set loose upon it.

Another class of white cells are called *natural killer cells* – (NK) cells – because they possess the ability to kill certain tumour cells or normal cells infected by virus. NK cells are inactive against 'free' virus particles, bacteria or parasites, and will attack only virus-infected cells. The NK cell releases specific proteins into the small space between its own membrane and that of its target. One of these proteins is *perforin*, a cylindrical molecule which is inserted into the target cell membrane, opening up a pore through it. Sodium ions flood into the target cell, drawing water behind them until it swells and eventually bursts.

Some types of white cell, especially *basophils* and *mast cells*, also secrete molecules that may produce an intense acute *inflammatory reaction* around the site of infection (called the acute phase). Basophils circulate in the bloodstream whereas mast cells are immobile and located in connective tissue, primarily the respiratory tract. When activated, these cells release preformed substances which mediate this response, the major one being *histamine*, and later synthesize in addition *prostaglandins* and *leukotrienes* from precursors stored in their membranes. Contact with them results in blood vessels dilating and becoming leaky so that plasma and white cells flood out, causing the swelling, redness and warmth that are characteristic of an inflammatory reaction.

Complement

The group of about twenty proteins that go by the collective name of the *complement system* form an important link between innate and adaptive immunity. Its overall function is to mediate phagocytosis and to control the inflammation. Some of the complement proteins are secreted by macrophages and normally circulate in the body fluids in inactive form. The proteins are related in the form of a cascade, so that when the first component is activated, a chain reaction is started in which the reaction products of each step in the sequence activate the next component. The cascade can be triggered by antibodies bound to the bacterial cell surface or directly by some bacterial products. Activation by either pathway generates peptides which opsonize microorganisms for uptake by phagocytes, attract phagocytes to sites of infection (*chemotaxis*), and increase the blood flow and permeability of capillaries to molecules in the blood plasma at the site of activation. The final product in the complement cascade is called the *membrane attack complex*.

This is a cylindrical array of molecules which, like perforin, can damage the cell membrane of some bacteria, viruses or other organisms which have induced the activation, by forming a pore through which sodium ions, and then water, flood into and finally burst the cell.

Inflammatory reaction

The various white cells and molecules that are involved in the processes of innate immunity are shown in Figure 54. This acute inflammatory reaction occurs not only at sites of infection but also around cuts, burns, grazes and other damaged areas, sealing them with an exudate of plasma, white cells and clotting factors that congeal into a protective scab. The reaction is accelerated by cytokines, glycoprotein molecules secreted by a variety of cells, among which in innate immunity macrophages are the most important. They recruit more cells which in turn release molecules that further enhance the reaction. This positive feedback ensures that the reaction, once initiated, proceeds very rapidly. There are other, inhibitory mechanisms which damp down the reaction and prevent it from spreading. However, these mechanisms can sometimes fail when the

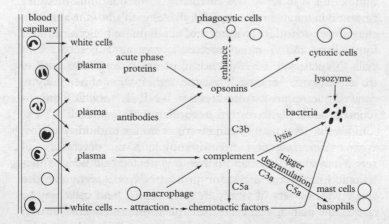

Figure 54. The mechanism of innate immunity

infection is persistent and then the inflammation can become chronic and serious damage to the tissues may result.

Cytokines induce the activation and synthesis of adhesion molecules on various blood and tissue cells, and these molecules in turn mediate the adhesion of neutrophils and monocytes at sites of infection, which is followed by migration, local accumulation, and activation of the inflammatory cells. These inflammatory cells serve to eliminate the bacteria, mainly by phagocytosis, followed by intracellular killing. Cytokines also induce fever and stimulate the synthesis of acute phase proteins. However, large amounts of cytokines or their uncontrolled production can be harmful and are responsible for some pathological manifestations of infection. The cytokines that mediate innate immunity include: *tumour necrosis factor* (TNF), *interleukin 1* (IL1), *interleukin 6* (IL6) and *chemokines* (IL8 family). All primarily function to initiate inflammatory reactions that protect against microorganisms.

TNF is a mediator of both innate and adaptive immunity and an important link between specific immune responses and acute inflammation. It induces fever and acts in the liver to increase synthesis of certain serum proteins. Like TNF, the main function of IL1 is to serve as a mediator of the host inflammatory response in innate immunity. IL1 shares with TNF the ability to cause fever, to induce synthesis of acute phase plasma proteins by the liver, and to initiate metabolic wasting. IL6 causes liver cells to synthesize several plasma proteins that contribute to the acute phase response. Thus the combination of hepatocyte-derived plasma proteins induced by TNF, IL1 and IL6 constitutes the acute phase response to inflammatory stimuli.

There are two main mechanisms of innate immunity against viruses. Viral infection directly stimulates the production of *type 1 interferon* (IFNα and IFNβ) by infected cells which functions to inhibit viral replication, whilst NK cells recognize and lyse a wide variety of virally infected cells. These cells may be one of the principal mechanisms of immunity against viruses early in the course of infection, before specific immune responses have developed. Type 1 interferon can enhance the ability of NK cells to lyse infected target cells. In addition, complement activation and phagocytosis serve to eliminate viruses from extracellular sites and from circulation.

Innate immune responses, including inflammatory reactions, begin within a few hours of invasion of the body by infectious organisms. This is much faster than the reactions of adaptive immunity, and these innate mechanisms are thus the body's sole means of defence in the early stages of infection – up to two to three weeks if the pathogen is being encountered for the first time, or two to three days if it has been met before, and the adaptive immune response has already been developed. It is to the mechanism of this response that we now turn.

ADAPTIVE IMMUNITY

The system of adaptive immunity has evolved in vertebrates to expand and refine the system of innate immunity by adding several new capacities. (1) An exquisitely precise and specific mechanism of immune recognition and response; (2) the ability to recognize practically any possible molecular structure; (3) a broader spectrum of effector mechanisms; (4) the ability to efficiently recognize and eliminate both extra- and intracellular microorganisms; (5) the ability to 'remember' each encounter with a microorganism so that subsequent encounters will result in a more efficient defence; and (6) the ability to amplify the protectective mechanisms of innate immunity and to focus them to the sites of antigen entry.

The adaptive immune response is initiated by the recognition of various molecular structures – *antigens* – on invading pathogenic microorganisms by receptors which already exist prior to an encounter of an organism with them. Recognition leads to activation of cells that have recognized antigens, and culminates in the development of effector mechanisms that mediate the biological effects of the response: *neutralization*, *destruction* and *elimination* of pathogens.

Immune recognition consists of the binding of antigens to corresponding pre-existing immune receptors on cells of the immune system. It is mediated by structurally extremely diverse cellular receptors that are synthesized by two main classes of circulatory cells, the T (derived from the thymus gland) and B (derived from bone marrow) lymphocytes which initiate adaptive immune responses. The working of the adaptive immune system depends on its potential to recognize the antigens of a vast number of various microorganisms even before it actually encounters them. Thus, the immune recognition system is one that anticipates what it should be looking for and what it must be potentially able to recognize in order to prevent death from infection.

During the effector phase of the adaptive immune response, B and T lymphocytes that have been activated by binding to an antigen perform functions which ultimately lead to destruction and elimination of antigens. Activated B lymphocytes differentiate into plasma cells which secrete antibodies, the effector molecules of *humoral immunity*. Some activated T cells differentiate into cells capable of secreting *lymphokines* which activate phagocytes to kill intracellular bacteria or stimulate the inflammatory response in general, and also promote differentiation of B cells into antibody-secreting plasma cells. Other activated T cells directly lyse cells infected with viruses. Both types of T cells represent important effector cells in *cell-mediated immunity*.

B Lymphocytes: producers of antibodies

An *immunoglobulin (Ig)* is a protein synthesized in an organism exclusively by B-lymphocytes. The Ig molecule can be expressed on the surface of B lymphocytes when it serves as an *antigen receptor*, or it may be secreted by cells in the blood plasma derived from antigen-activated B lymphocytes during an immune response to a substance foreign to that organism – i.e. the antigen. In the latter case, the secreted Ig molecules are called *antibodies*. The secreted antibody is *specific* simply because it is virtually identical to the original receptor molecule expressed on the cell membrane of B lymphoctyes which have recognized and bound the antigen.

An antibody binds to and *precipitates* (if the antigen is soluble) or *agglutinates* (if the antigen is particulate) only the substance that elicited its production. Macromolecules which are able to induce antibody synthesis – that is, serve as antigens – by B lymphocytes include proteins, glycoproteins, complex polysaccharides, phospholipids and nucleic acids. The specificity of an antibody is directed towards a particular site on the macromolecule, called the *antigenic determinant*. Macromolecules typically contain multiple determinants. Molecules below a certain size are not able to elicit antibody formation unless they are attached to a macromolecular carrier such as protein. Such small molecules which by themselves cannot elicit antibody formation but are able to react with antibodies like antigens are called *haptens*. When attached to macromolecules they are called *haptenic determinants*.

The structure of immunoglobulins

The basic structure of all immunoglobulin molecules is a unit consisting of two identical *light* or *L* polypeptide chains and two identical *heavy* or *H* polypeptide chains, linked together by disulphide bonds. Thus its unit structure may be written as L_2H_2 (figure 55). There are two distinct types of L chains called *lambda* (λ) and *kappa* (κ), and five distinct classes of H chains, *gamma* (γ), *mu* (μ), *alpha* (α), *delta* (δ), and *epsilon* (ε). Since the class of an immunoglobulin molecule is determined by its heavy chain, there are five classes of immunoglobulin molecules, called IgG, IgM, IgA, IgD and IgE, respectively. The commonest is IgG which accounts for 70 – 75 per cent of the total immunoglobulin pool in normal human blood plasma, while IgM accounts for approximately 10 per cent.

Ig chains are genetically related, grouped into the *Ig super-family*. Members of this superfamily contain repeating homologous units, about 110 amino acid residues long, which fold independently in a common globular motif called an immunoglobulin (Ig) domain. The Ig domain contains amino-acid sequences that permit the polypeptide to assume a globular tertiary structure containing an intrachain disulphide-

Figure 55. *The structure of IgG*

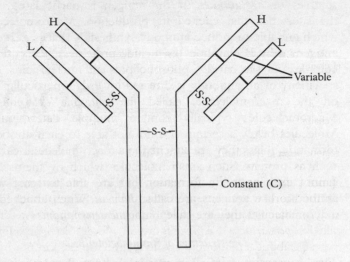

bonded loop. Many other proteins of importance in the immune system, such as antigen receptors on T cells, MHC molecules, CD4 and CD8 molecules, among others, contain regions that use the same folding motif and show structural relatedness to immunoglobulin amino acid sequence. Proteins belonging to this hyperfamily are evolutionarily related, structurally homologous and are therefore thought to have evolved from a common ancestral form.

The most striking peculiarity of Ig H and L polypeptide chains is that they are amazingly diverse, i.e. they all differ in amino acid sequences. This extraordinary diversity, whose generation will be explained later, accounts for the huge receptor repertoire for immune recognition, and the extraordinary specificity of antibodies for antigens because each amino acid difference may produce a different specificity in antigen binding. The amino acid sequence diversity, however, is not evenly distributed along the entire polypeptide chain, but is confined to the amino-terminal *V*, i.e. *variable*, regions consisting of about 110 amino acid residues, while the remainder of the

Figure 56. Hypervariable regions in V regions (domains) of L and H Ig chains

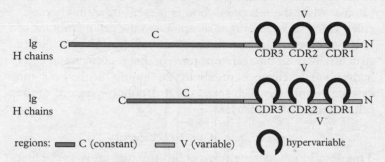

chains represent the more conserved *C*, i.e. *constant*, regions. Furthermore, within the amino-terminal V regions of both Ig H and L chains there are three highly diverse short stretches called *hypervariable regions* (Figure 56). These three segments are each about ten amino acids long, and are held in place by more conserved *framework regions*. Within the framework regions, certain amino acid residues and certain structural features are very highly conserved.

Since a basic unit of Ig molecules consists of two identical H chains and two identical L chains covelently linked to each other, and because the antigen-binding site is formed by interaction of V-regions of each H and L chain, an antibody usually contains two antigen-binding sites per one molecule. In forming an *antigen-binding site* the three hypervariable regions of each light chain and the three hypervariable regions of each heavy chain are brought together in three-dimensional space, while different amino acid sequences produce different antigen-binding sites with different specificities. Because these sequences are thought to form a surface complementary to the three-dimensional surface of a bound antigen, the hypervariable regions are also called *complementarity-determining regions (CDRs)*.

The constant region of the H chains carries out various effector tasks, such as the binding of complement, binding to various receptors on different cells, etc. All heavy chains may be expressed in one of two molecular forms that differ in amino acid sequence on the carboxyterminal end. The *secretory form* of the H chain is found in secreted antibody molecules in blood plasma, while the *membrane form* is present in antibody molecules which serve as antigen receptors on the cell membrane of B lymphocytes. The membrane form of the H chain differs structurally from the secretory form in that it contains an extra hydrophobic sequence of about 26 amino acids near the carboxy terminus, which spans the hydrophobic regions of the cell membrane lipid bilayer.

Generation of antibody diversity

This description of immunoglobulin structure answers the first of the questions we asked – how does an antibody bind to an antigen? – but leaves us to approach the much more problematic second question, of how specific antibodies are synthesized. In this respect it is important to remember one vital difference between antibodies and enzymes; normally antibodies obtained by injecting a foreign macromolecular antigen into an animal, such as human albumin into a rabbit for instance, are not simply made up of a single type of antibody but are a mix of various antibodies directed against various antigenic determinants of the given protein molecule. This mix of molecules all of which are antibodies against the same protein antigen is an intrinsic feature of the immune response. However, a single cell produces only one type of antibody directed against one specific antigenic determinant. It is different cells which produce the different kinds of antibody molecules directed against different antigenic determinants of a macromolecular antigen. Any normal antibody to a protein antigen in the blood plasma is thus a mix of antibody molecules secreted from many different clones of antibody-producing cells and is called a *polyclonal* antiserum.

During the 1940s and 1950s, even before the mechanism of protein synthesis was known, two rival theories competed to account for the rapid response of the immune system in producing antibodies capable of specifically recognizing novel molecules. The one which won out is called *clonal selection theory*, and is now generally accepted as one of the most important paradigms in immunology. Every adult individual contains many millions of different *clones* of lymphocytes, each clone having arisen from a single precursor cell prior to and independent of exposure to antigen. Each individual B or T lymphocyte belonging to a particular clone is genetically programmed to encode a receptor specific for a particular ligand. Thus cells constituting each clone have identical antigen receptors, which are different from the receptors on the cells of all other clones. In an antibody response, the antigen binds to and therefore *selects* a specific already pre-existing clone of B cells by attaching to their receptors, and thus activates them. Activated B cells proliferate and differentiate either into effector *plasma cells* or into *memory cells*. Plasma cells produce and secrete large amounts of the receptor molecule which when secreted constitutes the antibody. Memory cells retain the capacity to respond to the same antigen if encountered again. Their responses to the second and subsequent exposures to the same antigen are usually more rapid, larger, and often more specific than the first, or primary, immune response to that antigen. The latter property of adaptive immunity is called, for somewhat dubious reasons, *immunological memory*.

It is estimated that each individual mammalian immune response can recognize and discriminate at least a billion distinct molecular structures because it contains a similar number of differing clones of B lymphocytes and is therefore able to produce that number of different types of specific antibody molecules directed against virtually any foreign determinant with which it is or will be presented. The structural diversity of these molecules is much greater than that of all other molecules in the vertebrate body put together. This very impressive diversity of receptors – that is, antibodies – allows the immune system to recognize practically any possible molecular structure, thus assuring protection even against microorganisms that have developed strategies to avoid immune defence.

But how can such diversity be produced? As we have already seen, the basis of antibody specificity lies in the amino acid sequences of the variable V regions of the H and L chains. Each antibody chain is actually encoded by at least two genes, one variable V and the other constant C. In addition there are genes for joining segments (J genes) and, in the heavy chain, some so-called multiple diversity (D) gene segments. This

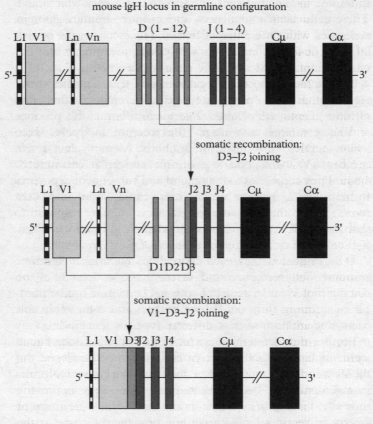

mouse IgH locus in germline configuration

somatic recombination:
D3–J2 joining

somatic recombination:
V1–D3–J2 joining

rearranged functional gene for IgμH

Figure 57. Rearrangement of gene segments (DNA recombination) in mouse IgH locus during B cell development

makes three main mechanisms of diversity possible: the presence of multiple genes for many possible V chains; mutations occurring during development to increase the number further; and recombination of gene segments to add yet more variety (Figure 57).

(1) The *multitude of V gene segments*. V regions of both the H and L chains are encoded by numerous V genes which have different nucleotide sequences and produce immunoglobulin molecules with different specificities. For instance, the mouse IgH genetic locus contains 250–1000 V gene segments and the κ L site contains at least 250 V gene segments.

(2) the *combinatorial association* of different V, D and J gene segments has a large potential for generating a great number of various receptor specificities. The maximum possible number of VDJ combinations, which form the third hypervariable region (CDR3) of V regions of Ig chains, is the product of the number of V, D and J gene segments at each locus. For instance, the random combinatorial joining of V, D and J gene segments at the mouse Ig H gene location can generate 10 000–40 000 diverse VDJ combinations (i.e. 250–1000 V × 12 D × 4J) which form the CDR3 of the V domain of H chain. Thus, every clone of B cells and their progenies express a unique combination of V, D and J gene segments. Because of combinatorial diversity, immunoglobulin molecules show the greatest diversity at the junctions of V and C regions in both Ig H and Ig L chains. These junctions form the CDR3, which is the most important portion of antibody molecules for specific antigen binding.

Furthermore, even the same set of germ-line V, D and J gene segments can generate different amino acid sequences at the VJ, VD, and DJ junctions. For instance, imprecise DNA rearrangement can occur because nucleotide sequences at the ends of V, J and D gene segments can each recombine at any of several nucleotides in the germ-line sequence. As long as the recombination does not generate nonfunctional DNA, different nucleotide and, subsequently, amino acid sequences can arise. Further, nucleotides coding for amino acids at the VD, DJ, VJ junctions can be added to the genetic coding sequences. This addition of new nucleotides is a random process probably mediated by an enzyme.

(3) Point mutations in V gene segments of H and L chains during so-called maturation of the antibody response can result in changes of antigen-binding specificities, and thus represent an important potential for generating additional diversity of antibodies.

Finally, (4) in addition to these mechanisms operating at the level of Ig genes, the *combinatorial pairing* of V regions of different H and different L chains which jointly define the specificity of the antigen-binding site of an antibody molecule also increases diversity. For instance, combinatorial pairing of V regions of 1000 different H chains with those of 1000 different L chains could theoretically produce 10^6 different antigen-binding sites. Thus *combinatorial pairing of chains* serves to yet further multiply the diversity generated for each chain.

Antibody production

B lymphocytes which bear specific immunoglobulin molecules as receptors on their surfaces are normally slow to divide but, having bound a particular antigen, they may start to proliferate and differentiate into plasma cells which synthesize and secrete large amounts of virtually the same molecule as the original receptor, but which is now called the antibody. Thus within a few days of the *first* exposure to a protein antigen, for instance, antibody molecules will begin to be produced by an injected animal. During this *primary response*, the first antibody molecules to be synthesized belong to the IgM class, which have a relatively high molecular weight – around 1 000 000. But before long the amount of IgM being produced decreases and a different class of immunoglobulin, IgG, with a molecular weight around 150 000, begins to be synthesized in greater amounts. This does not alter the V domain of the H chains, and thus does not change the specificity of antibodies for antigen, which remains the same. During the *secondary response*, which follows a second and subsequent exposure to the same antigen and is characterized by a more rapid and larger antibody response, it is IgG antibodies which are predominantly produced. The average affinity of these antibodies produced in a secondary response may be higher than in primary response due to *affinity maturation*, as a result of somatic mutations and selective preferential activation by antigen of those B cells whose receptors have increased affinity for that antigen as a result of mutations in the V regions of L and H Ig chains.

As explained earlier, B lymphocytes possess receptors capable of recognizing antigenic determinants on macromolecules such as proteins, glycoproteins, complex polysaccharides, phospholipids and nucleic acids, and are also able to produce antibodies against them when appropriately activated. However, protein antigens cannot induce appropriate activation of B lymphocytes in the absence of so-called helper T_H lymphocytes because B cells require two distinct types of signals for their proliferation and differentiation into plasma cells producing anti-protein antibodies. One type of signal is provided by the protein antigen itself which binds to the receptors on specific B lymphocytes. The second is provided by helper T_H cells and their secreted products.

During the effector phase of the antibody response, B lympocytes that have been activated by the antigen proliferate and differentiate into plasma cells which secrete antibodies, the effector molecules of *humoral immunity*. Secreted antibodies have either the same or even higher affinity for antigen binding as the original receptors. Secreted IgM, IgG and IgA antibodies neutralize viruses and toxins produced by bacteria, eliminate bacteria by enhancing their phagocytosis by blood neutrophils and mononuclear phagocytes, and activate the complement system which participates in the lysis and phagocytosis of bacteria.

T-LYMPHOCYTES – HELPERS AND KILLERS

The T cell receptor – TCR

Just as do B lymphocytes, T lymphocytes also express a huge number of structurally highly diverse antigen receptors on their membranes. These receptors, called TCR (T cell receptors), share a common underlying design and are quite similar in basic structure to the immunoglobulin receptors on B lymphocytes. They consist of polypeptide chains which also belong to the Ig gene superfamily because they contain repeating homologous units which fold independently in a common globular motif called an immunoglobulin domain.

The majority of TCRs consist of two linked polypeptide chains, α and β, with V (variable) and C (constant) regions. The amino terminal V regions, which participate in the formation of the antigen-binding site, are highly variable in terms of amino acid sequences that determine the specificity of interaction of an antigen-binding site with its particular ligand. In V regions of both the α and β chains there are three highly diverse *hypervariable regions*. The antigen-building site of the receptor is formed by interaction of V regions from the two polypeptide chains. The remainder of the chains represent the more conserved C regions. The genetic sites for the TCRs contain multiple V, D, J and C gene segments. The TCR genes in the earliest T lymphoctye precursors are in the nonfunctional configuration, which is characterized by the spatial separation of gene segments. During maturation of T lymphocytes in the thymus, the gene segments are rearranged in a defined order and placed close to one another. This process of DNA rearrangement results in the formation of functional TCR genes and is a prerequisite to TCR gene expression. The basic sequence and processes which mediate TCR gene rearrangement are quite similar to that for immunoglobulin, and the even greater structural diversity (perhaps up to ten billion possible molecules) is generated in a rather similar way.

In contrast to B lymphocytes, which are capable of recognizing an extremely large number of various antigenic determinants of different categories of macromolecules which can be either free in soluble form or cell-bound, T lymphocytes recognize only protein antigens and then only if the proteins or peptides are bound to a special class of molecules, the *Major Histocompatibility Complex* (MHC) on the surfaces of other target cells. Recognition of both the MHC molecule and the associated protein or peptide is mediated by a single receptor molecule which specifically interacts both with MHC-associated peptides and with the MHC molecules themselves.

MHC molecules

Class I and class II MHC molecules are expressed on the cell membrane and are responsible for immunological rejection of transplanted kidneys, hearts and other organs. The basic physiological function of these two main types of MHC molecules, which are fundamentally similar in structure and belong to the Ig gene superfamily, is however not to cause problems for transplantation surgeons, but to mediate recognition of protein antigens by T cells. Therefore the general antibody response to a protein is controlled by the presence or absence of MHC molecules that bind to and make available fragments of that protein to T cells. A second means by which MHC genes can influence immune responses to particular protein antigens is through the role of MHC molecules in shaping the receptor repertoire of T cells during their differentiation in the thymus.

Genes which encode class I and class II MHC molecules are by far the most varied in the genome of all species which have been studied. Thus each member of a species may contain a different MHC profile. With so many possible MHC molecules available within the gene pool of a species, bacterial peptides which can escape binding to MHC molecules in at least some individuals in the population are unlikely to emerge by mutations.

Class I MHC molecules are present on virtually all cells with nuclei, serving to protect against viral infection. In contrast, class II molecules are normally expressed only on cells called APCs (*antigen-presenting cells*), which include B lymphocytes, mononuclear phagocytes and some others. Cytokines can modulate the rate of transcription of class I and class II genes in a wide variety of cell types. This is an important amplification mechanism for T cell immune recognition function which already has found practical application in medicine.

CD4 and CD8 T lymphocytes

Because viruses and bacteria present quite different challenges to the immune system both in terms of recognition and appropriate responses, two parallel immune recognition modes have evolved among T cells. Peptides derived from intracellularly synthesized proteins, such as virus proteins, are presented in association with class I MHC molecules to a distinct subset of T lymphocytes which express CD8 molecule on their surface and are mostly cytolytic T_C lymphocytes. These cells are also called MHC class I-restricted cells. On the other hand, proteins from the external environment, such as toxins produced by invading bacteria, are presented in association with class II MHC molecules to another distinct subset of T cells which express the CD4 molecule on their surface and are mostly helper T_H cells. These cells are also called MHC class II-restricted cells.

CD4 and CD8 T lymphoctye surface molecules belonging to the Ig gene superfamily are thus expressed on two mutually exclusive subsets of mature T cells. Through their interaction with invariant domains of MHC class II and class I molecules, these molecules play an important role of co-receptors in antigen recognition by T cells, and shape the different functions and use of class I versus class II MHC molecules. CD4, in addition to its role as co-receptor in immune recognition, is a receptor for human immunodeficiency virus (HIV).

The differential antigen recognition by T lymphocytes makes good biological sense. It is clearly advantageous for an organism that its T lymphocytes capable of killing the cells, i.e. the CD8 MHC class I-restricted T_C lymphocytes, first recognize virus proteins in association with MHC class I molecules on the surface of virus-infected cells, and thereafter kill these cells in which foreign virus proteins are synthesized. This is highly desirable because eliminating such virus-infected cells before the virus in them has succeeded to start replicating essentially contributes to the eradication of the virus from the infected organism. However, it would make little biological sense to generate killer T_C lymphocytes against macrophages which, for instance, have taken up toxins produced by invading bacteria. Rather, they should, by presenting processed toxin proteins in association with MHC class II molecules, induce and activate CD4 MHC class II-restricted helper T_H cells which are capable of helping B lymphocytes in efficiently producing specific anti-toxin antibodies able to neutralize the toxins.

Signalling by lymphokines

In the activation phase of the adaptive immune response, a central role is played by a set of MHC class II cells which are precursors of helper T_H lymphocytes. These cells are involved in virtually all aspects of the adaptive immune response by initiating both humoral and cell-mediated immunity. Helper T_H lymphocytes activate other cells of the immune system directly or by means of chemical signals – the *lymphokines* (Figure 58). If depleted of such helper T cells, an individual is unable to mount an efficient immunological defence against any microorganisms, and is condemned to death, as dramatically manifested in patients in the final stages of AIDS.

Following antigen recognition, antigen-activated T_H cells first express IL2 receptors and produce IL2 molecules. The lymphokine binds to the IL2 receptor to induce cell proliferation, and may act either on cells in the immediate vicinity or feed back on itself to affect the cell that synthesized it. T_H cells proliferate and differentiate into two subsets of functional helper T_H lymphocytes, T_H1 and T_H2, which are characterized by differential lymphokine production.

(1) *Activated T_H1 lymphocytes* produce lymphokines which have three broad types of function. First, they provide essential signals for growth and development of functional killer *cytolytic T_C lymphocytes* which are the principal defence mechanisms of adaptive immunity against viruses and may be important in the immune destruction of tumours. Their most important role is probably elimination of virus-infected cells. They produce the membrane pore-forming protein called perforin, a series of enzymes called *serine esterases*, and protein toxins which all are

Figure 58. The role in the immune response of the main lymphokines produced by helper T_H lymphocytes

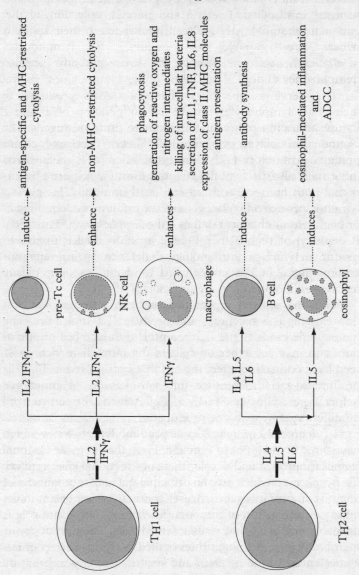

involved in T_C-mediated killing and lysis of target cells. Secondly, they mobilize macrophages to efficiently kill phagocytosed intracellular bacteria and tumour cells. Finally they stimulate the growth of NK cells and increase their ability to lyse target cells.

(2) *Activated T_H2 lymphocytes* predominantly produce lymphokines IL4, IL5 and IL6. These lymphokines, as well as IL2, serve as obligatory chemical signals produced by helper T_H lymphocytes during the cooperation of T and B lymphocytes in synthesizing antibody to protein antigens. This cooperation is necessary because protein antigens cannot induce activation of B cells in the absence of T_H helper lymphocytes, since B lymphocytes require two distinct types of signals for their proliferation and differentiation. One type of signal is provided by the protein antigen, which interacts with immunoglobulin receptor on specific B lymphocytes. The second type of signal is provided by helper T_H lymphocytes and their lymphokines. It initiates and stimulates growth and differentiation of antigen-activated B lymphocytes into plasma cells producing antibody.

Does this rather complicated way of making antibodies to protein antigens make any biological sense? Proteins are likely to be the most common source of antigens, and any protein molecule can be seen as a huge conglomerate of potentially antigenic determinants, for each of which there may be preformed specific receptors in each individual immune system. Therefore, even though a bacterium might confront the immune system with perhaps one million potential antigenic determinants, the immune system is probably able to generate clones of B lymphocytes to recognize all of them. But the implication of this is also that inherent in any antibody response to protein antigens is antigenic noise, a paralysing degree of potentially possible chaotic and disordered polyclonal activation. Therefore, a large number of potential antigenic determinants in proteins to which the immune system is constantly exposed should never be allowed to be recognized and allowed to induce antibody synthesis. It appears that the immune system solves this problem by having discriminating filters mediated mainly by antigen presenting APCs, including B lymphocytes. These cells maintain order in recognition by eliminating the inessential noise and by focusing the attention of the immune system on a manageably small part of the antigenic universe which it encounters. Thus one could envisage that APCs, by processing and presenting protein antigens, serve as a filter and a lens: a filter that eliminates the molecular noise, and a lens that focuses the attention of lymphocyte receptors on particular molecular signals.

SELF– NON-SELF DISCRIMINATION

At the beginning of this chapter we stated that 'at the heart of the immune response lies a capacity of the immune system to distinguish "self" from "non-self"' but we have not really explained what such a mysterious-sounding statement might mean in practice. The issue is simple enough. How can one prevent an immune system which is so marvellously sensitive from generating antibodies to an organism's own proteins – all of course containing potential antigens? This is not a trivial question. Failure to distinguish self from non-self results in a variety of auto-immune diseases, whilst enthusiastic recognition of non-self results in the rejection problem which plagues transplant operations.

Of course, any system that randomly generates a huge number of receptors to provide recognition of all possible shapes cannot avoid self-recognition. The potential for autoreactivity exists in all individuals because genes encoding T and B lymphocyte receptors that might recognize self antigens are present in the genome. Therefore, the need to have the capacity to distinguish 'self' from 'non-self' is a problem created by receptor diversity which each individual immune system must solve by learning. Thus, tolerance to self must be individually learned during development rather than being genetically programmed and inherited.

Self–non-self discrimination, also called *self-tolerance*, is the

Figure 59. T and B cell cooperation in the production of antibodies to protein antigens

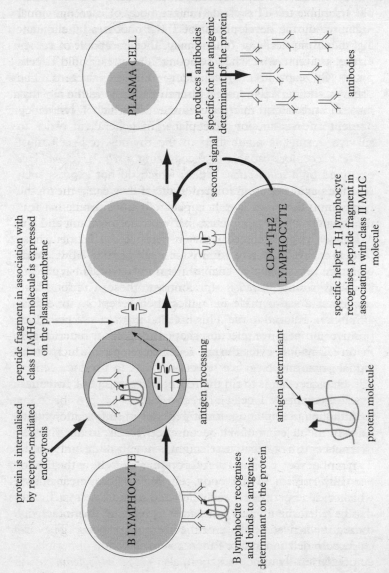

PLASMA CELL

produces antibodies specific for the antigenic determinant pf the protein

antibodies

peptide fragment in association with class II MHC molecule is expressed on the plasma membrane

second signal

CD4+TH2 LYMPHOCYTE

specific helper TH lymphocyte recognises peptide fragment in association with class II MHC molecule

antigen processing

antigen determinant

protein molecule

protein is internalised by receptor-mediated endocytosis

B LYMPHOCYTE

B lymphocyte recognises and binds to antigenic determinant on the protein

fundamental property of the immune system primarily carried out by a class of lymphocytes known as $CD4^+T_H$. In principle, tolerance to self is created in the body of each individual organism during development of T lymphocytes by eliminating, either physically or functionally, T cells capable of recognizing self antigens, while selecting and preserving T cells potentially capable of recognizing non-self antigens. The primary site for self–non-self discrimination is within the thymus, the specialized microenvironment in which T cell development and maturation take place. In general, in order to survive during its maturation in the thymus, a T cell must express receptors capable of recognizing a *self MHC molecule* occupied by a *self peptide*. T cells which do not express such receptors at all are programmed to die; if they can perform the recognition they *may* live. It is generally considered that it is the strength of the interaction between the receptor and the MHC-self peptide complex that is decisive. All T cells whose receptors have moderate affinity for a self peptide-MHC complex receive a signal that enables them to survive. This process is called *positive selection*. In contrast, those T cells which encounter a self-peptide for which their receptors have high affinity are induced to die. This is called *negative selection*. Both positive and negative selection shape the receptor repertoire of mature T lymphocytes during T cell development in an individual organism. Negative selection exists to produce central self tolerance, that is to rid the receptor repertoire of potentially self-recognizing T cells.

Although presentation of self peptides in sufficient amounts leads to the deletion of self-recognizing T cells in the thymus, the mechanism of negative selection is not absolute and mature T lymphocytes capable of responding to self antigens are normally present in the body. Further control mechanisms which induce peripheral self tolerance in autoreactive T cells that have left the thymus must exist to regulate their reactivity during the activation phase of the immune response. They include both deletion and/or functional inactivation of peripheral autoreactive T lymphocytes.

Deletion of self-recognizing clones of B cells is also required for preservation of self tolerance. Clones specific for membrane-bound self antigens present at high concentrations in the bone marrow are deleted before they become functionally competent, most probably at the stage in their development when they express only the IgM form of their antibody receptors. However, such deletion seems to play only a relatively minor role in B cell tolerance since many B cell clones capable of recognizing self antigens and producing antibodies to them are present in normal individuals.

Control mechanisms maintaining self tolerance may thus be interpreted as a series of interconnected safety valves built into the immune system and comprising a multifactorial interaction of several types of molecules and cells. Only if there is a simultaneous disruption of the majority of these mechanisms would autoaggression in the form of autoimmune diseases occur.

CHAPTER 14

The Unity of Biochemistry

In what has gone before, we have tried to describe the behaviour, in chemical and physical terms, of the biological unit called 'the cell'. But when speaking of this cell, we have in fact nearly always been referring to, not the cell of any one of the millions of different species of living things that inhabit the earth, but the cell of one particular group of animals: the mammals. Although on rare occasions we have digressed into the fields of plant and microbial biochemistry, such trespasses have nearly always been brief, and then normally intended to illustrate a point about the mammalian cell – even though many of the basic biochemical studies were originally made not on mammals at all but on less complexly organized creatures such as bacteria.

Indeed, we are not even being very fair in extending ourselves as far as even a description of 'the mammalian cell'. Most laboratory studies of mammals, and hence the experiments we have been describing, are done on only a very small number of different species. The commonest are rats and guinea-pigs, rabbits and hamsters – and then mainly males; very much less frequently domestic and farm animals such as sheep, cows and pigs are studied; and sometimes birds such as chickens (my own favoured organism) and pigeons. Always, attempts are made to relate this biochemistry of laboratory animals back to humans. But these species form only a very small fraction of the widely differing groups of known vertebrates. And when we come to plants, insects, or microorganisms, the number of whose known species overwhelmingly outnumbers the total of the vertebrates, such biochemical studies seem to fade into insignificance.

Yet we have confidently described 'the cell', its enzymes, metabolism, and behaviour almost as if unaware of the existence of so many other different sorts of cell remaining to be examined. Have we not been grotesquely over-confident in doing so?

Fortunately, we are fairly secure in saying no. The differences in appearance and overall behaviour between humans and yeast are indeed vast. Yet it is astonishingly, almost at first sight unbelievably, the case that most of the chemicals that compose the two are practically identical; almost without exception, their proteins are made of the same twenty amino acids, their nucleic acids of the same four purine and pyrimidine bases, their carbohydrates of the same or similar sugars. And even where slight differences may be pinned down between, say, the amino acid sequences of individual enzymes, the metabolic pathways catalysed by these same enzymes remain, in yeast and humans, identical over large regions. The pathway of glucose breakdown, studied originally in yeast as the fermentation of sugar to alcohol, was subsequently found to be identical with the route by which the muscle cell converted glucose to pyruvic acid. The only difference lay in the subsequent fate of the pyruvic acid.

Similarly, it is possible to purify an enzyme from one organism and use it without apparent difficulty to study a reaction in another, quite different one. Ribosomes from a microorganism and cytoplasm from a rabbit or duck will contentedly collaborate to synthesize protein. Even though the biochemical behaviour of only a minute percentage of the different forms of life on earth have been examined, one may still predict with a good chance of success that the general conclusions may be extrapolated to cover all other forms as well. The basic mechanisms of carbohydrate, fat, and protein catabolism and synthesis are the same in all forms of life now existing. This is a fact at first sight so unexpected and so surprising when one thinks of the manifold differences of form which life manifests, that it demands almost a positive effort of will to accept it. (This similarity is one of the reasons why those advocates of 'animal rights' who argue that no human benefit can possibly come from the study of the biochemistry, physiology or pharmacology of non-human animals are simply wrong.* On the other hand it is also the reason why for *some*, though by no means all problems, it is not necessary to work with mammals or even vertebrates but one can and should turn to creatures with less well-developed nervous systems.)

Its implications become all the more startling when one comes to consider the undoubted biochemical differences that do exist between different life-forms. The most profound of these differences lies in the sources from which the organism obtains the energy it requires in order to remain alive. Animals, fungi, viruses, and most bacteria rely on the existence of pre-formed organic compounds. We have dwelt at some length on the way in which animals burn glucose or fats to carbon dioxide and use the energy so released to synthesize ATP. Deprived of such preformed energy sources an animal rapidly wastes away and dies. In the presence of glucose and certain essential amino acids and vitamins, it can synthesize all the thousands of other chemicals it requires. Many bacteria are less demanding; they can survive on simpler 2- or 3-carbon organic acids, and some of them do not need amino acids but can make their own by transamination reactions with ammonia. Yet even these rather cleverer bacteria and fungi can normally exist as well (or better) in the presence of glucose and amino acids as they do in their absence. Most of them have the enzymic ability to deal with these substances just as animals can, even if the bacteria can do without if times get hard. They are just rather less specialized than the animals, and can therefore live rougher.

A second major difference lies between those organisms which require oxygen to act as an 'energy-sink' and oxidize their foodstuffs, and those 'anaerobic' bacteria which either never use oxygen, or can do without it at a pinch. Yet even such

---

* Of course this isn't the only type of argument that is deployed. Another claim is that, precisely because other animals are *similar* to humans, then they should have comparable rights. Many animal activists seem to want it both ways. Whilst I agree that we as humans have duties towards non-humans precisely *because* we are human, my ultimate loyalty remains to my species.

differences are more apparent than real. Those microorganisms which obtain their energy entirely by fermentation do so by pathways of metabolism similar to the routes taken in animals for the initial steps of glucose breakdown. The difference lies only in the fact that the fermentative organisms are unable to complete the process and oxidize the end-products of fermentation, and instead resort to a variety of tricks to extort the maximum energy from their excreta before finally discarding them as alcohol, acetaldehyde, lactic acid, or other essentially half-digested substances. In this case the animals possess oxidative abilities that the fermentative microorganisms do not. But their basic biochemistry is not greatly different, only more efficient.

A different class of anaerobes are those which can do without oxygen, not because they do not oxidize their substrates, but because they find an alternative hydrogen acceptor and ultimate 'energy-sink'. Typically, they use either sulphate or nitrate as their acceptors; the nitrate-reducing bacteria, for example, convert nitric acid, HNO_3, to nitrous acid, HNO_2, and in doing so gain an atom of oxygen to which they can pass electrons and reduce to water. But the most fascinating thing about these nitrate-reducing bacteria, seemingly operating on such different principles from the animal world, is that the electron-transport pathways by which they pass their hydrogen to nitric acid are identical with those of the animals which use the oxygen of the air. The substrate yields its hydrogen to a dehydrogenase which passes it on to intermediate hydrogen carriers which are cytochromes. The only difference is that, in the final stage, instead of using cytochrome oxidase to take the hydrogen from the cytochrome to oxygen, the nitrate-reducing bacteria oxidize their cytochromes with a different enzyme, nitrate reductase, which converts nitric to nitrous acid and water (Figure 60).

AUTOTROPHES AND HETEROTROPHES

The real dividing line in the biochemical world comes between those groups we have just discussed, all of which depend on the

Figure 60. The difference between humans and nitrate-reducing bacteria

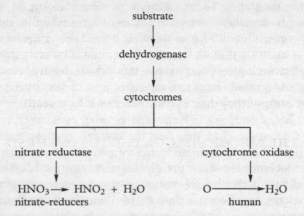

substrate

↓

dehydrogenase

↓

cytochromes

nitrate reductase   cytochrome oxidase

↓   ↓

$HNO_3 \rightarrow HNO_2 + H_2O$ $O \longrightarrow H_2O$

nitrate-reducers   human

existence of preformed organic compounds for survival, and those living things which can make all their own organic substances from simple inorganic materials. Such organisms are called *autotrophes*, indicating that they are self-sufficient by comparison with the *heterotrophes*, like yeast and humans, who have to be cushioned by the existence of sugars and amino acids against the harsh realities of the inorganic world.

Autotrophes do not obtain their energy by burning ready-made fuel, but cast about to find an alternative energy-source. The paradigm case, of course, is the green plant, which avoids the heterotrophe's dilemma by trapping the light energy pouring down on to the earth from the sun, and using it to 'fix' carbon dioxide as organic carbon and, ultimately, to synthesize sugars and starches, a process known as photosynthesis. The 'higher' green plant is not alone in performing photosynthesis; both algae and a group of photosynthetic bacteria can perform similar reactions. Another group of autotrophes (the *chemoautotrophes* as opposed to 'photo-autotrophes') use the energy latent in certain inorganic chemicals for carbon dioxide fixation instead – by the oxidation, for example, of sulphur, ammonia, or hydrogen. These biochemically fascinating chemo-autotrophes, many of which can be found living in the extreme environments on the edge of volcanoes or close to thermal vents in the depths of the ocean, now seem to form an evolutionary backwater, an essay in chemical versatility that did not quite come off, although in the early period of the evolution of life on earth they were, as we will see, much more important. Most research attention has, not unnaturally, been focused on the photosynthesizing organisms because of the present-day importance of their role in the totality of life on earth.

THE MECHANISM OF PHOTOSYNTHESIS

The problems faced by the plant are in essence identical to those of the animal. They may be summed up as the need to trap energy and obtain a source of primary building blocks so they may carry out the biochemical synthesis of more complex molecules. Both forms of life obtain their precursors from essentially the same source – breakdown of glucose to CO_2 and H_2O via glycolysis and the citric acid cycle – the routes and the enzymes used are very similar. Both plants and animals use ATP as their energy reservoir and in both it is formed by linking its synthesis to the passage of electrons along a chain of carriers. The real difference lies in the source of the glucose and of the energy that is trapped in this way.

The ultimate source of the energy plants use is sunlight. During daylight hours, light energy is trapped by a process involving chemiosmosis and a system of electron carriers very similar to that used in the oxidative reactions described in Chapter 7. But whereas in the process described previously, energy for the synthesis of ATP is released by using oxygen to oxidize $NADH_2$ to NAD and release CO_2 and H_2O, in photosynthesis the process runs, as it were, in reverse; electrons are passed from H_2O to NADP, *releasing* oxygen from the water and providing reducing power in the form of $NADPH_2$. The essential feature of this *light reaction* in plants is that a stepwise electron transfer pathway is involved in which the splitting of H_2O and release of O_2 is linked to ATP formation by a proton-pumping, chemiosmotic mechanism. The subsequent fixation

353

of CO_2 to synthesize glucose utilizes the ATP and $NADPH_2$ synthesized during this light reaction, and is called the *dark reaction* because, if the plant already has adequate supplies of ATP and $NADPH_2$, it can indeed take place in the dark. The double set of light and dark reactions together comprise photosynthesis and normally, during daylight hours, both reactions

$$6CO_2 + 6H_2O + \text{light energy} \longrightarrow C_6H_{12}O_6 + 6O_2$$

take place together. The overall reaction then can be written:

In this way plants provide themselves, and the heterotrophic world as well, with the only ultimate source of organic compounds at present available in the world outside the chemist's synthetic test-tubes. The total 'fixing' of CO_2 that occurs by photosynthesis is prodigious, providing as much as 112×10^{11} tons of organic carbon a year. At the same time it continually renews the $O2$ of the atmosphere and removes the CO_2 accumulated during respiration, providing a turnover so rapid that every molecule of CO_2 in the atmosphere will be incorporated into glucose by photosynthesis on average once every 200 years. Between them, the photosynthetic mechanisms of the plant and the respiratory system of heterotrophes provide for the regular revolution of the 'carbon cycle' which takes so prominent a place in every biology textbook. The combination of increased release of CO_2 by burning fossil fuels and reduction in the fixation of CO_2 because of destruction of vast tracts of the plant cover of the earth is today resulting in the steady increase in atmospheric CO_2, changing the balance and causing the atmospheric warming described as the 'greenhouse effect'.

The analogies between the photosynthetic light reaction and hydrogen transport in animals are not merely chemical or mechanistic, for photosynthesis takes place in a subcellular structure called the *chloroplast* which has a striking structural similarity to the mitochondrion (Figure 61 and Plate 8). Like the cristae of the mitochondrion, the grana and lamellae of the chloroplast provide the sites for ATP synthesis and hydrogen

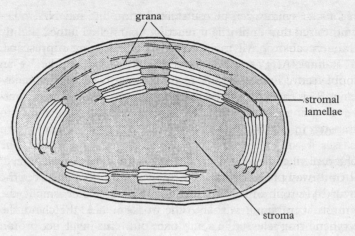

Figure 61. A chloroplast

(electron) transport, and have a lipoprotein structure similar to that already described for the mitochondrion.

In the trapping of light energy, the first and most critical of the steps of photosynthesis, the pigment *chlorophyll* (which gives plants their green colour) is all important. Although chlorophyll is by no means the only photosynthetic pigment, it is the only essential one. The molecule has a hydrophobic hydrocarbon tail by which it becomes firmly embedded in the chloroplast lamellae. The polar head part of the molecule is in fact very similar in design to that of the haem of the cytochromes and haemoglobin (see page 152). Like haem, it consists of a linked series of four carbon-and-nitrogen containing rings ('pyrrole rings') joined together to form a sort of doughnut with a hole in the middle. This hole is filled in haem by the metal iron; in chlorophyll on the other hand the jam in the doughnut is made of magnesium. The ring structures contain a series of alternating double and single bonds, and the absorption of a given small amount of light (a quantum) of a particular wavelength causes a sort of vibration, or resonance around these bonds. Because of the close packing and stable orientation of the pigment molecules within the lamellae, this resonance energy can be transferred from one pigment molecule to another until it is eventually channelled into a slightly different chlorophyll molecule from which it cannot escape. This final energy-trapping type of chlorophyll can receive an input from as many as 300 of the standard chlorophyll molecules. The energy from the light is thus very highly concentrated at a single site, giving the second molecule the ability to transfer an electron to a non-pigment receptor which in turn passes it, via an intermediate set of carriers, to NADP. The chlorophyll at the central site, which has thus become oxidized, is converted to its original state by accepting electrons from the hydroxyl group of water. Thus the water molecule is split, the protons combining with NADP to form $NADPH_2$ and the oxygen being released in a sequence of events involving proton pumping via chemiosmosis which bears, once again, a remarkable similarity to the respiratory electron transport chain and ATP synthetic mechanism:

$$2H_2O \; + \; \text{oxidized chlorophyll} \longrightarrow$$
$$\text{reduced chlorophyll} + O_2 \; + 4(H)^+$$

$$4(H)^+ \; + \; 2NADP \longrightarrow 2NADPH_2$$

In fact the situation is made more complex because there are two different such reaction sites each receiving energy from two different pigment systems; the flow of electrons from H_2O to $NADPH_2$ involves the cooperation of both pigment systems.

Let us return to the so-called dark reaction of photosynthesis. In it, both the $NADPH_2$ and the ATP formed in the light reactions are consumed in the fixation of CO_2. The fixation reactions were charted by Melvin Calvin and his co-workers in Berkeley, California in the 1950s and 1960s, with the use of radioactive CO_2. During these reactions, CO_2 is made to combine with a pentose (5-carbon) sugar containing two phosphate groups, ribulose bisphosphate, to give an unstable 6-carbon intermediate which breaks down to two molecules of the 3-carbon phosphoglyceric acid.

$$C_5 \quad + \quad C_1 \quad \longrightarrow \quad C_6 \quad \longrightarrow \quad 2C_3$$

| ribulose | carbon dioxide | | intermediate | | phosphoglyceric acid |

Phosphoglyceric acid lies (see page 167) on the well-mapped pathway of glucose metabolism; some of it can be used to manufacture fructose and glucose phosphates (see page 199) whilst some of the other molecules of phosphoglyceric acid are recombined through a maze of interlocking reactions to resynthesize the 1-phosphate sugar ribulose phosphate. ATP is used, finally, to rephosphorylate ribulose phosphate to ribulose bisphosphate, and the cycle can start up again. Most of the enzymes concerned are virtually identical to those of the pentose phosphate and glycolytic pathways that we have already discussed, the exception being that enzyme which actually initially fixes the CO_2, ribulose bisphosphate carboxylase, which is a large complex allosteric molecule with a molecular weight of around 300 000, subject to inhibition and activation by many different substances. Needless to say, this particular enzyme is closely concerned with the control of what became known as the Calvin cycle.

But the essential point to note is that, with the exception of the chlorophyll-containing apparatus responsible for the splitting of water and hence providing the primary energy source, all the reactions of photosynthesis, fixation of carbon dioxide, and synthesis of sugars follow pathways with which we are already familiar in the biochemistry of the animal cell. Once again, what at first sight appears to be a major difference in biochemical systems, between photosynthetic green plants and heterotrophic animals, is in fact more startling for its similarities.

COMPARATIVE BIOCHEMISTRY AND BIOCHEMICAL EVOLUTION

To argue in this way is not to deny the very real evidence of interesting biochemical distinctions between species – and even between individuals. Every human, with the exception of identical twins, is genetically unique. And because of the ways in which genes are differently expressed during development depending on experience and environmental differences, even identical twins are in some ways biochemically different. And it is possible to interpret such differences in terms of biochemical evolutionary processes. Multicellular humans are probably right to feel themselves more biochemically complex than anaerobic bacteria in that they can oxidize their food all the way to carbon dioxide and water, which demands more biochemical finesse and subtlety than the microorganism, which can live rougher but only tap off a small portion of the potential energy of glucose before being obliged to discard as refuse such energetically potent substances as alcohol or lactic acid. Similarly, very simple organisms contain in their cells neither nucleus nor mitochondria, whose presence in more complex ones makes an obvious contribution to the stability and efficiency of the cell.

An interesting example of such biochemical evolution in action comes from Ernest Baldwin's study of the mechanisms of nitrogen elimination. The problem faced by mammals in disposing of the nitrogen produced during the breakdown of protein is that this nitrogen exists in the form of ammonia, and ammonia is extremely poisonous even in very small quantities. Thus the animal, in order to avoid building up lethal quantities of ammonia, converts it instead into urea and excretes it in the urine. But not all animal species dispose of their ammonia in this manner. Bony fishes, for example, are content to excrete ammonia intact and without further conversions, whilst reptiles and birds instead produce not the soluble urea but the highly insoluble uric acid, which is excreted as solid nodules.

Why this difference in metabolism? Baldwin showed that it could be related to the availability of water to the different species. In fresh-water fishes, water is constantly available and continually diffuses into and out of the fish. Under these circumstances the ammonia formed will be washed out of the bloodstream as a very dilute solution and carried away into the surrounding water, without there being time for toxic concentrations to accumulate. Bony sea-fishes are in a similar position, complicated by the fact that the water in which they live contains salt at a higher concentration than that in the blood, and they therefore have to take steps against either too great an influx of these salts from the sea, or too great a loss of water into the sea. So they can only excrete part of their nitrogen as ammonia, but have to convert the remainder (about one-third) into other less poisonous substances.

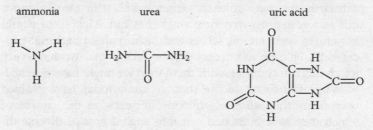

Land animals, though, are in a quite different position. They have to conserve water to avoid dying of desiccation. So they cannot afford to dilute ammonia down with large amounts of water before getting rid of it; instead, they turn it into the far less dangerous urea. The most interesting example of this adaptation is provided by the frog, which whilst a tadpole excretes nitrogen as ammonia, but, on changing into its adult, land-based form, also rearranges its internal biochemistry so as to make urea instead.

Finally, the birds and reptiles. The young of these species are hatched from eggs, but unlike the water-borne eggs of fish, the eggs of birds and reptiles are laid on dry land. In order to avoid complete drying out while waiting to hatch, they therefore have hard shells which are impermeable to water. But this impermeability means that they have no way at all of disposing of their waste nitrogen in solution. This clearly rules out making nitrogen into ammonia, and even disposing of it as urea would mean that concentrations of urea would begin to build up in solution towards the time when the egg is due to hatch, and these concentrations could be quite unpleasantly high (enough to give the embryo a rather bad headache, at the least). So instead, the ammonia is converted to the quite insoluble uric acid, which is harmless and can be disposed of, not as a solution, but as small solid nodules. The whole story provides a fascinating example of the operation of adaptive evolution at the biochemical level.

But despite such examples of biochemical evolution and modification, the biochemical composition and organization of all the forms of life now present on earth demonstrate an extraordinary unity quite at variance with their more obvious differences in gross structure and behaviour. Most of us would hesitate to compare ourselves with fish, typhoid bacteria, cancer cells, or even oak trees, yet the fact is that we have very much more in common with them than we might have guessed.

Why? The forms of life that we know today have evolved over many thousands of millions of years, in the course of which they have branched out into such a host of diverse directions that their outward resemblances are remote. Yet during the whole of this evolution, their biochemical forms have remained remarkably constant. The only convincing explanation for this is that these biochemical parameters were established and fixed before the species began to evolve along the different pathways that biologists have traced. Insofar as the biochemical forms are identical, it must be the case that all existing species had a common ancestor about whose external shape and form we can only guess but about whose biochemistry we may be certain that it was very similar to that of living things today. We can express this relationship thus:

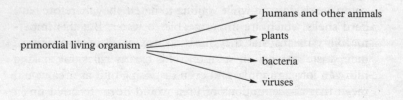

primordial living organism

humans and other animals
plants
bacteria
viruses

THE ORIGIN OF LIFE

It is such an analysis that has led biochemists into one of the most fascinating of their hunts – that for a convincing description of the origin of life on earth. If all current-day manifestations of life can be accounted for in strictly chemical and physical terms, as all biologists would maintain, it follows that it ought to be possible also to describe the *origins* of these present-day life forms in chemical and physical terms. The alternatives would be either to assume that the primordial ancestor of both beast and human was set up by some non-chemical and physical intervention and then left to go along under its own chemical and physical steam for ever afterwards; or to maintain that life in one form or another has always existed, for as long as the universe itself, and therefore had no need to arise specifically anywhere.

Neither of these possibilities is intellectually satisfying, nor are either of them at all necessary provided we can demonstrate a convincing way in which life *could* possibly have arisen without violating chemical and physical principles. If we can provide such an account, in broad terms, it does not matter unduly if some of the details later have to be amended in the light of further scientific advances; the important thing is to show that, even in our present-day limited scientific state, we can nonetheless provide an explanation which does not breach known scientific principles and is logically satisfactory. It is up to those who wish to mystify the nature of life then to criticize these hypotheses, if they are determined to try to show that life could not have arisen in the way that is proposed.

The principle of spontaneous generation of everything from barnacles to microorganisms has been widely believed throughout much of human history. It was the rigorous investigations of Pasteur in France in the middle of the nineteenth century that conclusively demonstrated that life as it now exists could not have arisen spontaneously from non-living matter. Every living thing, said Pasteur, has arisen from another living thing. Nor should this surprise us, for even the simplest of present-day living organisms are highly complex, highly improbable molecular structures, whose chance assembly from their elements would involve odds of such astronomic unlikelihood that we may regard it, for practical purposes, as impossible. The chemicals which compose present life-forms require to be synthesized by specifically catalysed reactions, and these specific catalysts are themselves the product of the living organism and cannot arise spontaneously. To seek for the origins of the complex of attributes that we regard as life today, we must assume that these attributes evolved only slowly over the 4.5 billion years of life's history on earth.

In order to provide some account of the way in which these attributes may have arisen, it is necessary to make plausible suggestions about the conditions that existed on earth at that remote time. The first rigorous and systematic attempts to do this were made more or less simultaneously in the 1930s by J. B. S. Haldane in Britain and A. I. Oparin in the Soviet Union and their explanations still provide the bedrock for modern theories. Before their work, difficulties had arisen because it was assumed that the earth's primitive atmosphere was largely oxygen and that the first organisms to evolve must have been capable of performing photosynthesis in order to trap energy and synthesize the organic substances they needed. Yet photosynthesis, as we have seen, is a highly complex process clearly only possible to already well-developed and highly-skilled organisms. This dilemma was resolved when Oparin was able to point out that the atmosphere of the primitive earth, far from being oxygen-rich, must have resembled that of the other planets, containing vast quantities of hydrogen, methane (marsh-gas, CH_4), ammonia, and carbon dioxide. The present-day atmosphere has replaced this primitive one precisely *because* of several billions of years of life, turning methane and carbon dioxide into organic chemicals, and, eventually, once photosynthesis evolved, releasing oxygen (it is the current greenhouse effect which is in danger of turning the clock back once more). The obvious fact that the presence of life changes the planet even while the nature of the planet constrains the forms of life that can evolve upon it, so that animate and inanimate nature constantly and dialectically interact, has been mystified in recent years into claims that the earth itself can be regarded as an organism, the so-called Gaia hypothesis of James Lovelock. Such a nebulous and potentially confusing concept is not however necessary in order to appreciate and work with the idea that it is not only species but also environments that evolve.

Today's debates about the origin of life are mainly about the conditions under which complex organic syntheses and the development of self-replicating molecules, notably the DNA-RNA-protein systems, could have occurred. There is a powerful school of thought amongst molecular biologists that life is *defined* by the capacity of accurate self-replication, and hence that only with the emergence of proto-genes, of RNA or DNA, could life be said to begin. As will become clear in the rest of this chapter, this is not a view I agree with. There is more to life than replication.

In the Oparin–Haldane model, the primitive earth could be pictured as containing huge warm oceans in which were dissolved a variety of salts derived from rocks, and over which hung an atmosphere of gases which would rapidly be lethal to any presently-living organism. Under these conditions, a number of organic compounds would have begun to be formed and scattered in solution throughout the sea. The formation of these compounds would have depended precisely on the reducing atmosphere and the steady influx of energy in terms of light and ultraviolet radiation from the sun, for in such circumstances CO_2, H_2O, CH_4, and NH_3 can react to give a mixture of products including amino acids, urea, and many other substances.

An interesting experimental verification of this idea was provided more than 20 years after Oparin and Haldane had first advanced it by Stanley Miller in America, who passed an electric charge through a mixture of H_2, CH_4 and NH_3 in a closed water-bath for periods of 20 hours or more. At the end of this time the products were analysed and found to contain more than eight different amino acids and seven monocarboxylic and dicarboxylic acids, all of which are amongst the basic building blocks of present-day organisms. Similar experiments have even been able to demonstrate the synthesis of ATP and of small proteins under completely non-biological conditions of this type. An alternative source for the origin of such organic chemicals is volcanic – Sydney Fox in the US has shown that in conditions mimicking those of the intense dry heat of volcanoes, peptide bond formation and the synthesis of proteinoid-type substances can occur.

Thus the primitive ocean must have steadily increased in organic content. The problem is how this dilute, evenly distributed solution could become concentrated enough to enable further reactions to occur. One possibility is provided by the fact that the surface of the rocks and clays of the beds of the shallow seas, containing iron, magnesium, and copper, would have provided binding and catalytic surfaces on which the organic substances would have begun to collect and to polymerize. As a result, short-chain peptides and nucleic acids, and possibly carbohydrates as well, would also have begun to accumulate, both bound on to mineral surfaces and free in solution in the seas. Some, such as the chemist Derek Cairns-Smith, have argued that this clayey organic mix, by binding the organic molecules and providing catalytic surfaces, was the real start of life.

By contrast, in the Oparin–Haldane model the critical stage in the process may have taken place in the oceans. It had been known for many years that solutions containing large molecules, such as the polymers of amino acids or carbohydrates, have a remarkable tendency to break up into small droplets containing the polymers in concentrated form, leaving the surrounding water comparatively free of dissolved substances. Salts and low-molecular-weight organic substances present in the solution also tend to be sucked into these droplets together with the polymers. This phenomenon is called coacervation, and has a perfectly logical, though somewhat involved, explanation in physical laws. Such coacervate drops may be formed from mixtures containing, for example, gelatin or gum arabic, and have been extensively studied in the branch of physics known as colloid science.

Oparin argued that, in the primitive oceans containing polymeric organic compounds, just such coacervate drops would have begun to be formed – the organic material would all have tended to coalesce into small, highly concentrated droplets. Within the droplets, the different compounds which had collected would have begun to interact with one another because of their new proximity. Coacervation as an idea is now part only of the history of science, but it provides a clue to a more modern version of the concept. The key lies in the physical properties of lipids. Put a drop of oil into water and, depending on the amount, it either forms a thin skin on the surface or curls up into a lipid droplet or micelle. The droplet may either be entirely lipid, or consist of a lipid membrane surrounding an internal solution. In essence a lipid micelle of this sort is a proto-cell. The internal constituents are concentrated within, and segregated from the outer world, by a lipid membrane. In that ancient half-living world we can imagine such droplets forming, some stably, some unstably, as the mix of chemicals within them dictated. For instance, reactions which resulted in falls in pH would tend to break the droplet up. Other droplets would continue to increase in size as they absorbed materials from outside until, beyond some optimum size, they would tend to split in two, each containing a more or less identical mix of the original components.

And so the process would have continued. Unstable droplets would have broken down and their organic material have become available once more for incorporation into stable ones. Stable droplets would have grown and divided. Within them, more and more complex polymers would have been formed. Metal ions acting as catalysts for favoured reactions, and co-enzymes such as nucleotides, would have become more active as they became bound to the peptide polymers which were the forerunners of proteins, thus forming proto-enzymes. It is at this point that the idea of the cell as a metabolic web, introduced in Chapter 11, becomes important. As the complexity of the chemical soup within the proto-cell increases, certain autocatalytic processes will develop and, as the mathematical modeller Stuart Kauffman pointed out, using chaos theory, the web becomes self-stablizing (Figure 62). The organic chemist J. P. R. Williams has added a fascinating gloss

Figure 62. An autocatalytic set. Food molecules (a, b, aa, bb) are built into a self-sustaining network. Reactions are represented by points connecting large molecules to their breakdown products; dotted lines indicate catalysis.

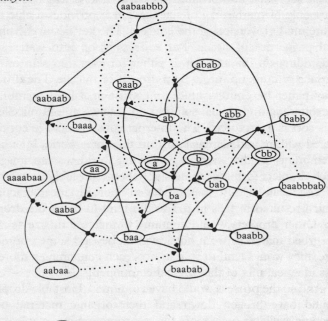

to this account. Early life forms, he argues, made use of relatively few of the chemical elements – carbon, hydrogen, oxygen and nitrogen, of course – but, as life evolved, so more and richer varieties of chemical reactions became exploited. Early on, phosphorus and sulphur chemistry, perhaps later the metal ions sodium and potassium, iron, calcium and magnesium. The evolution of life has to be seen also in terms of the evolution of the chemistry it exploits, and there have been sufficient examples of such complexity in earlier chapters for this idea to seem very attractive. Over the course of many hundreds of millions of years, the oceans would have become peopled with these stable, reproducing, primitive, semi-living droplets surrounded by membrane-like structures.

At some stage during this period in their development, the nucleic acids and proteins would have arisen as interdependent and mutually-synthesizing molecules, to form the forerunners of the DNA-RNA-protein complex which is today responsible for genetic transfer. Today's molecular biologists and bio-chemical geneticists working in the Crick tradition would regard this as the key event; for them the essence of life is accurate self-replication, the faithful copying of molecule by molecule, rather than the existence of cell-like structures. This replication property for them is not available to cells, nor complex organic mixes, nor even proteins, but resides in nucleic acids, and therefore what needs to be explained is how these self-replicating lengths of DNA or RNA can have begun. For them a system can only be described as living when it contains a nucleic acid-protein complex capable of precise self-replication and mutation. Clearly, a key feature of today's living organisms is their capacity accurately to reproduce themselves. When and how did this evolve? Is the present genetic code an evolved form from some primordial ancestral, simpler version? Because in some senses RNA is a simpler molecule than DNA, some molecular biologists have argued that the initial replicator was a single-stranded RNA. This argument, for a so-called 'RNA-world', has been given further credence by the fact that some viruses use RNA instead of DNA, that there are 'reverse transcriptase' enzymes which will synthesize DNA on RNA templates, and that there are examples of RNA acting enzymically – so-called ribozymes. Thus the versatility of RNA seems in some sense greater than that of either DNA or proteins. However, as Chapter 10 made abundantly clear, the idea that either DNA or RNA can 'self-replicate' without the involvement of a host of enzymes and cofactors is clearly misconceived, despite the currently fashionable ways of speaking about genes amongst some popular-science writers. And, again as the preceding chapters have made clear, synthesizing polymers is thermodynamically unfavourable. It requires a source of usable chemical energy. Thus the arrival of DNA or RNA replicators on the living scene must have been postponed yet further, until the development of such utilizable energy sources through autotrophic mechanisms.

As primitive cells began to accumulate in the seas, the availability of preformed organic substances must have steadily diminished. An evolutionary period must have arisen when there were not enough such molecules to go round. At this stage, those cells which could make use of inorganic energy sources such as hydrogen sulphide would have been favoured. For a brief period, such autotrophes must have flourished more extensively than they do today. It would seem likely that it was at this stage in evolution that the mechanism of photosynthesis developed. Once the photosynthetic and autotrophic organisms had evolved, though, the conditions of existence for other organisms must have changed for the better – oxygen would have begun to appear in the atmosphere and the stock of preformed organic material in the ocean have risen sharply once more. Ultimately, a self-regulating 'carbon cycle' between heterotrophes and autotrophes would have come into play, and the era of life as we know it today would have opened. Primitive autotrophes, with their special mechanisms of energy generation by way of proton pumping and ATP synthesis, may have been very similar to the mitochondria and chloroplasts found in today's organisms. It may have been during this period in life's evolution that some autotrophes became directly incorporated into the cellular structures of other organisms, making for more complex cells and ultimately more complex organisms. This would account for the puzzling residual amounts of DNA and RNA found in mitochondria and chloroplasts isolated from multicellular animals and plants (mentioned briefly in Chapter 4). This thesis, of the ultimate in symbiosis, has been developed extensively in recent years by the American evolutionary biochemist Lynn Margulis. Indeed Margulis goes further, arguing that *all* the subcellular components of modern eukaryotes are the absorbed symbionts derived from once free-living organisms – *proctista*, the basis, she believes, of all complex life.

Obviously, all such schematic accounts beg many questions. Some are complex chemical issues, such as that of the origins of the universal existence in living organisms, but not in non-living nature, of 'asymmetric' molecules like those of the amino acids or sugars (see Chapter 2). Nor are there yet very satisfactory accounts of the polymerization steps which produced proteins, nucleic acids, and so on from their more primitive ancestors. Some still maintain, like the astronomer Fred Hoyle, and even, in his more mischievous moods, Francis Crick, that the earth was 'seeded' with preformed nucleic acids present on comets. Indeed, since the development of space probes and the possibility of seeking life-forms on other planets, a whole new area – part experiment, part theory, called xenobiology – has grown up around the discussion of such ideas.

Other questions are more theoretical and philosophical. Just when, for example, in this evolution of living from non-living, can we be said to have stepped across the border between the two? What, in fact, is the definition of living as opposed to non-living? However, it is probably not very useful to try to draw a hard and fast line between the living and the non-living. Clearly some things – dogs, flowers, yeast cells – are alive. Others – such as molecules of salt, urea, or amino acids – are not. Between the two extremes lies an uncertain half-world filled with lipid micelles, pieces of nucleic acid, viruses, and some biochemical preparations like isolated mitochondria or nuclei. There is no hard and fast dividing line between living and non-living, any more than there is between a fertilized ovum in the womb and a full-grown adult, or between a raw and a hard-boiled egg. The two extremes are quite different, but the one is converted to the other by an infinite series of small steps, and it is only at the extremes that one can be very precise.

Such issues have raised heated debates at the various congresses on the origin of life that have been held over recent years. But they are debates which have been fought out within the framework laid down by physical and chemical theory and its applicability to complex systems. This is not to say that life reduces to 'mere' chemistry and physics. There are biological principles which express the organizing relationships between macromolecules, cells, and organisms, and which must include within them an understanding of historicity, for as the descriptions in this chapter have shown, biological systems have to be understood in temporal as well as molecular terms if their development and evolution are to become meaningful. Nothing in biology makes sense, said the great evolutionary biologist Theodosius Dobzhansky, except in the light of evolution. My version of this statement is broader: nothing in biology makes sense except in the light of history. Such principles, however, are materialist; to understand the existence and origins of life, and of humans, needs no recourse to principles outside those of the material world.

CHAPTER 15

Can Biochemistry Explain the World; Can Biochemistry Change the World?

In this book, we have moved rapidly across the field covered by modern biochemistry, separating the various stages in thought, experiment, and theory which characterize the biochemical approach to life, and at the same time showing how biochemists now believe that they can draw up a general balance-sheet which can account, in broad terms, for those aspects of living behaviour we can study. Aspects of many fundamental life processes can now be described in quite precise molecular terms, analogous to those in which the chemist writes equations for the reactions of simple acids and alkalis, or the physicist for the quantum energies of the electrons of reacting molecules. For many other qualities that go to make life we are far from being able to do this; our understanding is still too superficial. Furthermore, in many respects there are still fundamental theoretical problems in knowing what a 'complete' biochemical description of a living system would be.

There are some molecular biologists and biochemists who believe that a total description of the physics and chemistry of the cell would readily extrapolate to a total description of the organism. For example, the entire nucleotide sequences of several viruses and bacteria are now known. And in 1998 the complete sequence for a small nematode worm was published. But does such a specification say all there is to say about the properties of the worm? Or are there other things to say which, as I believe, can only be specified in terms of the history of the worm as an organism and its relationship to its environment? And if this is the case for worms, how much more will it be true for the ambitious programme to sequence the human genome which, at an estimated cost of some \$3 billion, is expected to be completed early in the coming century.

Those of today's molecular biologists who insist on a rigid genetic, molecular reductionism, a total explanation of the world in molecular terms – a view expressed most succinctly by James Watson ('in the last analysis, there are only atoms') – are, it seems to me, committing a philosophical and scientific error akin to that of the physicist Kelvin in the nineteenth century who argued that a 'complete' physics was only possible when all phenomena could be reduced to mechanical analogues – clockwork models. This mechanical molecular materialism, which underlies the crudity of the 'central dogma' of molecular biology discussed in Chapter 10, will need to give way to a much richer understanding of the need to interpret the phenomena of life at a series of levels, from the molecular to that of the population. No given level should ever be seen as fundamental or as static. Hence my preference for the term homeodynamics over homeostasis. All life has a history, a bio-chemical as well as an evolutionary and developmental history; the task of the biochemist becomes that of understanding living processes at just one of these levels, and of collaborating in the discovery of the translation rules that relate biochemistry on the one hand to physics and chemistry, on the other to physi-ology, psychology, ecology.

The preceding chapters have perhaps given the impression that modern biochemistry represents the inevitable conquering march of science out of an ignorant error-ridden past into the glorious light of today's understanding. Such an impression would be quite wrong. Certainly the advance of biochemistry ever since the 1960s, when the first edition of this book ap-peared, has been phenomenal, and we are still in the thick of this progress; there is no sign yet of slackening off. Many things only half-understood or mistakenly believed today will have to await a second and a third generation of researchers from now before they can be fully comprehended. Then those of us who have the temerity to publish our theories and concepts as those of the triumphant biochemistry of the opening years of the new millennium will either be forgotten in the inexorable advance of science or at best half-remembered as those whose insights were later verified or disproved. We should never be allowed to forget, in our enthusiasm for recent results and modern work, that today's biochemists and molecular biologists are building on the work of the chemists and physicists of the nineteenth century, and on those of the pioneers who created biochemistry early in the twentieth century as a science where none existed before. Too many graduate biochemists and post-doctoral researchers seem to work on the assumption that what wasn't published in the most recent weekly issues of *Nature*, *Science*, or one of the hundreds of specialist biochemical journals, is archival, and what happened before 1980 prehistoric. Only recently have there been some steps towards the creation of a history of biochemistry; more of us need to study it.

It is not only that we have to keep in mind how much is still not known, some of which is deliberately hinted at in the last few chapters; it is also that experiments, facts, descriptions which seem complete in one context may, with newer and greater understanding, need to be reshuffled and reinterpreted. Biochemistry has not yet been through the convulsions of the transition from Newtonian gravitational theory to Einsteinian relativity, that shook the foundations of physics in the early part of the century, though the trauma that molecular biology faced in coming to grips with the non-linearity of genes and the complexity of processing and regulatory mechanisms, described in Chapter 11, may come close. The mechanisms of control processes and the regulation of the cell, the intimate details of the biochemistry of cell structure, the functioning of the cell as part of the organism as a whole, the biochemical mechanisms of development, hormonal control, and memory – these are all problems which seem to us today as large and in many ways as difficult and intractable as the determination of protein structure or the elucidation of the citric acid cycle did to our biochemical elders.

All of these are soluble problems, granted support, time, and a theoretical approach which avoids an arid molecular reductionism. And such solutions are of importance not only to biochemistry itself, but outside too, for the practical application of results based on biochemical technology has come over the past thirty years to be of increasing importance. When the first edition of this book appeared in the 1960s, it was true to say that biochemistry was associated with relatively little derived technology. Biochemists were to be found in the fermentation and food industries (Pasteur was perhaps the first, and wine and cheese are but two of the important products of microbial fermentation), whilst pathology laboratories and the pharmaceutical industry absorbed the talents of others. Biochemists might believe we could explain the world, but few believed we could significantly change it by our science.

But that was all before the terms biotechnology and genetic engineering appeared in the dictionary, let alone on the lips of politicians and in the daily reports of quoted companies on the Stock Exchange. It is worth reminding ourselves just how far things have moved in the last thirty years. By the early 1970s it had become apparent that the new techniques in biochemistry, molecular biology and molecular genetics might make it possible to specifically alter the DNA at least of microorganisms so as to change the proteins that they produced. These tailor-made organisms might have a whole new range of properties – for instance, they might contain the enzymes required to degrade plastics, or they might contain the gene to produce such proteins as the hormone insulin – or they might indeed have wholly new disease characteristics. The potential excited many molecular biologists. But it also alarmed them to the extent that, led in the US by Paul Berg, a conference was convened at Asilomar, in California, at which many of the world's leading molecular biologists endorsed a proposed moratorium on experiments which might alter an organism's DNA. This self-denying research ordinance did not last more than a few years however. The pressures engendered by the thought of possible new scientific breakthroughs, and still more the commercial enthusiasm for the prospect of new products, led to a steady relaxation of the moratorium and the research guidelines imposed by the regulatory bodies which governments had set up, in the US, Britain, and many other countries, to police the new research.

Now, the pressures were all in the other direction. A rash of genetic engineering companies began to be floated on the Stock Exchange, many with science-fiction sounding names and distinguished university biochemists and molecular biologists on their boards, in their employ, or among their stockholders. Business-forecasting magazines estimated the total market for biotechnological products by the year 2000 as anything between $15 and $100 billion. Researchers with shares in the new companies became paper millionaires almost overnight; *Nature* began running a regular table charting the Stock Exchange fortunes of the leading biotechnology companies. The military, sensing the potential of the new technology in developing a new generation of chemical and biological weapons, began to move in. Politicians in Britain and elsewhere in Europe began to speak of a 'biotechnology gap' in research and development by comparison with the US and Japan. A tension developed between the scientific pressure to publish results and the commercial pressures to patent them. Universities themselves became more commercially minded and began to patent the discoveries of their researchers, while esoteric conflicts broke out, to be settled only after long and costly court battles, as to whether it was possible to patent engineered organisms, as 'new life-forms'. A new variety of philosopher, calling themselves 'bioethicists', began to speculate on the moral legitimacy of cloning, modifying and creating new life-forms. Would what had begun with bugs ultimately be possible for humans, offering prospects of a new eugenics?

By the mid-eighties a little more caution had begun to creep into the forecasting. Few of the biotechnology companies were actually making profits, and rather fewer new products had actually appeared in the market. Some of those that had were running into unforeseen difficulties. The share values of the biotechnology companies faltered.

By the beginning of the nineties rather more modest claims were being made for biotechnology, but some new products were now in more or less routine production, notably medical diagnostic kits made possible by the use of monoclonal antibodies, that is antibodies generated against a single antigenic determinant or epitope, and derived from a single clone. Some industrial processes made increasing use of modified microorganisms in fermentations associated with waste and sewage disposal, and also monoclonal antibodies for the large-scale purification of proteins and other macromolecules by affinity-binding methods and the use of enzymes, bound to surfaces or membranes, as catalytic agents.

The mix of disciplines and techniques which by the end of the 1990s together had been given the generic title of biotechnology or, more evocatively though misleadingly, 'genetic engineering' have been derived from many different sciences, not just biochemistry but also microbiology, genetics and molecular biology. In equal partnership with these are the methods and skills of chemical and process engineering, required to solve the problems of the scale-up from laboratory processes which are carried out in test-tubes in solutions of a few millilitres to the vast fermentation and extraction vessels required for the bulk production of microorganisms or proteins, which may be carried out in volumes of 200 000 litres or more.

For us the most relevant biochemical point is that the methodology now potentially exists for identifying and obtaining the DNA sequence for any one of the many hundreds of millions of naturally occurring proteins, and indeed, for artificially creating new proteins of specific sequence by synthesizing the requisite DNA for them. Advances in computer simulation of tertiary protein structures make it possible to predict the properties of a potential protein in an appropriate ionic environment from its primary sequence – and hence to design and synthesize artificial enzymes, tailor-made for specific reactions which may not occur in nature but might be of potential industrial or medical significance. Because DNA strands can use one another as templates and therefore can be copied identically, or cloned, DNA for proteins that exist only in minute quantities can itself be generated in bulk and can be used to produce indefinitely large quantities of the new protein.

The key to these possibilities lies in the assembly of

techniques which together have come to be classified under the general term genetic engineering. Genetic engineering essentially involves the insertion, deletion or modification of part of the genome of an existing organism. Such methods, originally applicable only to prokaryotic organisms, have become increasingly powerful and can now be used with laboratory species, ranging from nematode worms to mice. In 1997 history was made by a sheep called Dolly, born as a clone – that is, from cells whose own content of nuclear DNA had been replaced by that derived from an adult ewe. Despite the storm of ethical concerns that such cloning raised, clones of mice and cows soon followed. There is no scope in this book to discuss in detail either the methods or their actual and potential applications, or social issues raised by genetic engineering (sometimes called ELSI or ELSA – the Ethical, Legal and Social Implications or Aspects). A brief summary is however appropriate.

One of the first necessities is to catch your DNA. If you are intent on cloning an entire organism, you need all the nuclear DNA (and of course even this leaves independent the mitochondrial DNA, which is still contributed by the cell whose own nucleus has been removed and which is intended to serve as the vehicle for the clone). But most genetic engineering involves tinkering with specific portions of the genome, or even individual genes. Most individual gene-size DNA strands only exist in minute quantities, but the fact that specific DNA chains can be synthesized on a template of existing DNA means that in principle even small quantities can be accurately copied and thus produced in some bulk. This is the idea that lies behind the *polymerase chain reaction* (*PCR*), in which a strand of DNA is subject to sequential cycles of copying via the polymerase systems described in Chapter 10. PCR is used for a multitude of purposes these days, although its best-known popular role is in amplifying minute samples of DNA for DNA fingerprinting, for identifying tissue samples to confirm or deny parentage or the identity of victims or perpetrators of crime.

DNA strands can be merged, manipulated, have segments deleted or added by other technologies to produce so-called *recombinant* or *rDNA*, and the next task is to insert them into the genome of the organism whose genetic constitution is to be modified. The method essentially mimics a natural process – that by which a virus can subvert its host cell's nuclear mechanisms to start copying and manufacturing its own DNA. The procedure is therefore to incorporate the rDNA into the viral machinery (actually a virus-like structure called a *plasmid*) and inject it into the intended host cell, where it becomes in turn incorporated into the host DNA. Perhaps surprisingly, the procedure works, not always, but sufficiently often and accurately enough to make it commercially attractive. The result is that if the DNA represents a structural gene, the protein may begin to be expressed. If what is being manipulated is a promotor region, then it is also possible to affect the expression of a whole number of proteins. If the protein is an enzyme, it in turn may catalyse reactions which result in the accumulation of other interesting metabolites.

When these procedures were originally developed it was assumed that their major purpose would be to turn common microorganisms into biochemical factories for the bulk production of substances of medical or industrial use. It is easy therefore to see why the prospect of such techniques has generated such widespread interest, why some have seen them as an industrial panacea, the biochemical equivalent of the production of unlimited energy from seawater by fusion, and why others have viewed their potential with great concern. In reality however, there are both practical and theoretical problems which are likely to continue to limit the power of the new biotechnology and prevent it from becoming a biological perpetual motion machine. First, the production is not necessarily cheap. Second, even relatively 'simple' microorganisms are not clockwork machines, simply synthesizing whatever protein they are instructed to make by DNA. There are complex regulatory processes, some of which we touched upon in Chapter 10, which all organisms have evolved to protect their genome and limit its subversion from outside, whether by viruses as in nature or now by overconfident biotechnologists. Inserted DNA strands may be expelled, excised or inactivated by other mechanisms after a few cycles of replication. Third, bulk-produced 'alien' proteins may themselves be modified by other cellular mechanisms. The large-scale purification of such proteins from the bacterial broth in which they are generated is also not a simple matter. Problems of this type have limited the successful manufacture and application even of the most apparently straightforward piece of bacterial genetic engineering, that of the simple protein hormone insulin, even though it has been in commercial production and licensed for human use since the early 1980s.

The difficulties of rDNA technology in single-celled microorganisms become all the more daunting when the same techniques are applied to multicellular organisms. Plants are easier to manipulate than animals because in many cases they can reproduce asexually as well as sexually; a few carrot cells in a broth can regenerate an entire carrot and DNA inserted into the carrot cells will be copied along with the cloned cells. Sexual reproduction makes the transgenerational perpetuation of artificially inserted DNA much more problematic. But even prior to the question of reproduction, the fact that in organisms other than bacteria the DNA is not free in the cell but is confined to the nucleus, and only a tiny percentage of the total DNA of the organism is actually transcribed into protein, the rest of it being involved in a regulatory or even unknown (junk' or 'selfish' DNA – see Chapter 10) functions, makes the genetic engineering of complex organisms much harder.

The great range of editing and regulatory mechanisms that are built into the protein synthetic machinery in eukaryotic and especially multicellular organisms and which are described in Chapter 11 make the results of genetic manipulation far harder to predict. The two major obstacles to such prediction are development and sex. Thus although it is possible to alter or delete a specific gene in a germ cell or fertilized egg, one cannot be sure that an altered protein will be expressed at all, or whether during cell division and specialization during development it will be switched off or otherwise inactivated. Furthermore, the multitude of regulatory processes that multicellular organisms possess means that the effect of adding or subtracting a particular gene may well be compensated for by changing the expression of others. Thus during the 1990s techniques were developed for breeding mice (so-called *transgenics*) with particular genes either deleted ('knocked out') or added ('knocked in'). Many such 'constructed' animals carry lethal mutations and do not survive. But many which do survive, in the absence of a gene responsible for coding for a protein believed to be of major importance, seem perfectly normal. What has happened is that during development other genes have been brought into play to compensate for the missing one. This homeodynamic plasticity is a vital aspect of life processes, however inconvenient it may be for genetic engineers. Furthermore, the genetic shuffling that occurs during sexual reproduction may well mean that inserted genes are edited out or deleted, so progeny do not breed true. Hence of course the attention paid to cloning, which may bypass this problem. But before one gets too carried away with the prospects, as offered by the cloners of Dolly, of flocks of sheep or herds of cattle producing medically relevant biochemicals in their milk as a result of creating transgenes, one should remember that the successful cloning of a single sheep required several hundred failed attempts. No couple who have been through the agonizing cycles of attempting to conceive by IVF need to be reminded of the problems.

Nonetheless speculations about human gene therapy to eliminate 'unwanted' genes or to introduce 'desired' ones are rife. Leaving aside for the moment the ethics of such suggestions, the complexity of the biology involved should again be emphasized. There is, as we have made clear in Chapters 10 and 11, no simple one-for-one relationship between a length of DNA and a particular phenotype. Genes do not function in isolation but as part of a complex and exquisitely controlled orchestration of processes involving the interactions of hundreds of thousands of different such DNA segments with each other and the environment in which they are expressed in a developmental process occurring in many millions of cells.

The subtlety of the control processes that have evolved during thousands of millions of years to limit the possible disruption of this orchestra by outside 'noise' or by genetic 'mistakes' is still much beyond the reach of biochemistry to understand, predict or, in most cases, reliably to manipulate. Most of the phenotypes that are of interest because they represent disease traits that might be eliminated by genetic manipulation are the consequence not of the presence or absence of a single gene but of the interplay of many. Certainly there are some devastating diseases which occur as the result of a change in a single gene; phenylketonuria (page 136) is one, in which a single gene is affected, resulting in the absence of a key enzyme in the metabolism of phenylalanine, and therefore the accumulation in cells and blood of high concentrations of other intermediates which have toxic effects and ultimately result in irreversible brain damage. Sickle cell anaemia, in which a single DNA base is mutated, resulting in an altered haemoglobin structure (page 39) is another. So are cystic fibrosis, Duchenne muscular dystrophy, hypercholesterolaemia, Lesch-Nyhan syndrome and many hundred other single gene disorders ranging from the relatively common (one in every few hundred births) to the extremely rare in which only a few cases are known in the world. Quite apart from the human pain and suffering such conditions imply, all are of great interest to medical doctors, geneticists and biochemists because of what can be learned about cellular metabolic processes from them. Until the late 1990s, the regulations concerning human gene therapy distinguished between so-called somatic gene therapy, in which attempts are made to rectify faulty genes in an individual – child or adult's – body, and germ-line therapy in which the manipulation was attempted in the gametes themselves, with the implication that the genetic change would be transmitted to subsequent generations, as in the transgenic mice. Somatic gene therapy was to be permitted, germ-line therapy with the implication of tampering with future generations was not. However, where somatic gene therapy has been tried, even for diseases with known genetic causes such as cystic fibrosis, it has not been successful. It is hard to get the gene into place, and hard to ensure that it functions appropriately when there. Germ-line therapy would be easier, it is argued, because it requires only manipulating a single egg, and this could be done during what are now almost routine IVF procedures. As this book is being written, the pressure to drop both this moral barrier and that against human cloning is mounting, and may prove irresistible to those who see such high technology medicine as the path to the future.

But we should never forget that the overwhelmingly largest causes of human suffering and disease in the world are not these genetic conditions at all, but still the great killers of poverty and malnutrition, environmentally-precipitated cancers and heart disease, mental distress and disorder. In the understanding and treatment of such conditions, biochemistry, genetics and their associated technologies have but a small part to play.

This does not dissolve away the ethical problems associated with the question of whether genetic technologies could and should be employed in the detection, treatment or possible elimination of such conditions. For instance, the diagnostic kits based on monoclonals make early detection of a potential genetic disorder possible in the embryo or foetus by amnio-centesis or the newer technique of chorionic villus screening; even earlier detection is available in the case of human eggs removed from the mother for *in vitro* fertilization by the newest techniques. Following such detection the prospective mother may today be offered the choice of abortion; soon however it may be possible instead to opt for a genetic therapy by which the 'faulty' gene will be replaced by a 'corrected' one in the embryo or foetus, or even eliminated in the germ-line itself. One may question how great a priority should be given to such research and treatment when so many wanted babies die throughout the world each year through avoidable and simple diseases of poverty, malnutrition and dysentery. Nonetheless, the technological drive towards the development of such therapies in the medical-industrial complexes of Europe, the US, Australia and Japan seems irreversible. Nor can one ignore the military dimension at a time when chemical warfare using an earlier generation of biochemically-generated substances has been successfully waged by the Iraqis in the gulf war of the late 1980s, and active research is continuing in many countries.

This is the background against which the present debate about the ethics of genetic engineering, *in vitro* fertilization and human embryo experimentation must be judged, although much of the sound and fury of the debate has not echoed around such questions of practicality and priority, where in my view it ought to be primarily located, but instead in more mystical ways around the assumed 'rights' of the embryo and its status within established religious doctrine. Even the associated debate about the genetic manipulation of non-human organisms has been focused as much about the claimed 'species rights' of organisms not to be manipulated (as if plant and animal breeders had not been doing this for millennia) as about the potentially hazardous consequences of the release of novel organisms into an environment of whose physical and biological interconnections we have such a limited and inadequate understanding.

But we are here merely skating on the surface of these themes, to indicate something of the ways in which in the last few years biochemistry has moved beyond its esoteric status within the laboratory and become a science whose deep moral, political and technological implications cannot be wished away. Whilst they must not be sensationalized, they should equally not be trivialized or 'left to the experts' – whether such experts are biologists, philosophers or politicians. Nor are we in a stationary phase; the breakthroughs likely in the next few years portend far more than this: potentially revolutionary medical technology; the application of the understanding of brain mechanisms as techniques of mind control by precisely tailored molecules. Such possibilities have been actively canvassed.

Will they be beneficial? Above all, this depends on the shape of the society which permits or denies their application.

It is not a question of biochemists, or molecular biologists, as demi-gods in white coats threatening, or sensitively refraining from threatening, the future of the rest of humanity; that future depends more on the structure of our society than on the structure of our science, though each helps determine the other. However, biochemistry is 'special' knowledge, to which access is limited. Those who have it, have also the responsibility of spreading this knowledge, helping it to serve, not oppress the people, and learning the limitations of their own special knowledge in the process.

To approach the questions which form the title of this final chapter: biochemistry by itself is not enough to explain the world or even the human portion thereof. However, it is an essential part of the totality of that explanation which is the goal of true science. And biochemistry and its associated sciences increasingly have the power to change the world in important ways. How, and in whose interests, is a question for all of us.

A Note on Further Reading

There is a huge range of biology, biochemistry, molecular and cell biology textbooks available, and everybody is likely to have their own favourites. Most of the big academic publishers have biochemistry course textbooks, presumably adopted by enough US colleges to make them profitable. Amongst those readily available and authoritative are those listed below. I cite the most recent editions, but most publishers regularly update. Readers will note a certain lack of originality about the titles!

B. Alberts, D. Bray, J. Lewis, M. Raff, K. Roberts and J. D. Watson, *Molecular Biology of the Cell* (3rd edition, Garland, 1994)

S. R. Bolsover, J. S. Hyams, S. Jones, E. A. Shephard and H. A. White, *From Genes to Cells* (Wiley/Liss, 1997)

M. K. Campbell, *Biochemistry* (Harcourt Brace, 1999)

A. L. Lehninger, D. L. Nelson and M. M. Cox, *Principles of Biochemistry* (Worth, 1993)

R. H. Garnett and C. M. Grisham, *Biochemistry* (2nd edition, Saunders/Harcourt Brace, 1999)

H. R. Horton, L. A. Moran, R. S. Ochs, J. D. Rawn and K. G. Scrimgeour, *Principles of Biochemistry* (2nd edition, Prentice Hall, 1996)

L. Stryer, *Biochemistry* (4th edition, Freeman, 1995)

D. D. Voet and V. G. Voet, *Biochemistry* (2nd edition, Wiley, 1995)

Then there are of course the Open University course texts, of which the most recent and relevant is S327, *Living Processes*.

For the interested reader wanting to keep abreast of current developments in biochemistry and related sciences, a subscription to *New Scientist* and/or *Scientific American* is indispensable. More detailed reviews are carried in the Elsevier monthly journal *Trends in Biochemical Science*.

Index